Tucholsky Wagner Zola Scott Sydow Fonatne Freud Schlegel
Turgenev Wallace
Twain Walther von der Vogelweide Fouqué Friedrich II. von Preußen
Weber Freiligrath Frey
Fechner Fichte Weiße Rose von Fallersleben Kant Ernst Richthofen Frommel
Hölderlin
Engels Fielding Eichendorff Tacitus Dumas
Fehrs Faber Flaubert
Eliasberg Ebner Eschenbach
Feuerbach Maximilian I. von Habsburg Fock Eliot Zweig
Ewald Vergil
Goethe London
Mendelssohn Balzac Shakespeare Elisabeth von Österreich
Lichtenberg Rathenau Dostojewski Ganghofer
Trackl Stevenson Doyle Gjellerup
Mommsen Tolstoi Hambruch
Thoma Lenz Hanrieder Droste-Hülshoff
Dach Verne von Arnim Hägele Hauff Humboldt
Reuter
Karrillon Garschin Rousseau Hagen Hauptmann Gautier
Damaschke Defoe Hebbel Baudelaire
Descartes
Wolfram von Eschenbach Schopenhauer Hegel Kussmaul Herder
Darwin Dickens Rilke George
Bronner Melville Grimm Jerome
Campe Horváth Aristoteles Bebel Proust
Bismarck Vigny Barlach Voltaire Federer Herodot
Gengenbach Heine
Storm Casanova Tersteegen Grillparzer Georgy
Lessing Gilm
Chamberlain Langbein Gryphius
Brentano Lafontaine
Strachwitz Claudius Schiller Kralik Iffland Sokrates
Katharina II. von Rußland Bellamy Schilling
Gerstäcker Raabe Gibbon Tschechow
Löns Hesse Hoffmann Gogol Wilde Gleim Vulpius
Luther Heym Hofmannsthal Klee Hölty Morgenstern
Roth Heyse Klopstock Kleist Goedicke
Luxemburg Puschkin Homer Mörike
Machiavelli La Roche Horaz Musil
Navarra Aurel Musset Kierkegaard Kraft Kraus
Nestroy Marie de France Lamprecht Kind Kirchhoff Hugo Moltke
Nietzsche Nansen Laotse Ipsen Liebknecht
Marx Ringelnatz
von Ossietzky Lassalle Gorki Klett Leibniz
May vom Stein Lawrence Irving
Petalozzi Knigge
Platon
Sachs Pückler Michelangelo Kock Kafka
Poe Liebermann
de Sade Praetorius Mistral Korolenko Zetkin

The Library of Work and Play: Gardening and Farming.

Ellen Eddy Shaw

Imprint

This book is part of the TREDITION CLASSICS series.

Author: Ellen Eddy Shaw
Cover design: toepferschumann, Berlin (Germany)

Publisher: tredition GmbH, Hamburg (Germany)
ISBN: 978-3-8491-9112-2

www.tredition.com
www.tredition.de

The Library of Work and Play

GARDENING AND FARMING

BY ELLEN EDDY SHAW

GARDEN CITY NEW YORK
DOUBLEDAY, PAGE & COMPANY
1911

The Library of Work and Play

GARDENING AND FARMING

BY

ELLEN EDDY SHAW

1911

THIS BOOK IS DEDICATED TO THE REAL BOYS OF THIS REAL CLUB AND TO THE GIRLS WHO ARE JUST AS GOOD AS BOYS

CONTENTS

PART II.—THE CHIEF'S GARDEN TALKS

ILLUSTRATIONS

PART I—THE GARDEN CLUB

I

THE GARDEN CLUB

The door opened. A gust of wind and rain literally swept five boys, wet and breathless, into the room. The man at the big oak table in front of a huge open fire looked up, smiled, and said, "Off with your duds, boys! Bar the door securely, Jay, for it's a wild night. Throw a fresh log on the fire, Albert. And all line up."

For a few moments the big cheerful room seemed full of wriggling, twisting boys as great coats were pulled off and hung up carefully on pegs at the far end of the room. It was a rule here at The Chief's home that things should always be shipshape. Then the "line up" came. This was a little ceremony the boys always went through, having dropped into it quite of their own accord. They formed a line in front of the open fire with backs to it and faces toward the man. Then they solemnly saluted in military style. At this The Chief arose, saluted, and by a wave of the hand assigned each boy to his place at the table.

This little group of boys had formed itself naturally into a club. It met with The Chief every Saturday night. He was really no chief, this big, boy-loving man who had come to spend a while in this little country village, to rest and to write. The boys had named him The Chief because, as Albert said, "He could lead any tribe and tame any savage." At this Albert always laughed for he himself had been called a savage so many times he almost believed he was one.

The boys dropped into their places. Jay, or the "Little Chief" as the boys called him, sat opposite the Big Chief at the end of the table and right in front of the fire. He was slim and tall and light of foot. He could run faster, throw farther, and play better than any other boy in the village. He always led, he never bullied, he played fair, so the other boys always followed.

Albert, Jay's brother, big and heavy and as full of mischief as he could hold, took his place at The Chief's right hand. Albert called this his place of honour although knowing full well that he was there so The Chief might have him within reach. Next to Albert came George, frank-faced and bashful, sturdy and loyal. Opposite him red-headed, stubby Peter sat always on the edge of his chair, always with a bit of a smile on his face, never talking much, always agreeing good-naturedly. Beside Peter and at The Chief's left was Jack, who wriggled constantly like a young eel.

After the boys were seated and quiet, The Chief pushed back his work, a plan of his summer's garden, leaned back in his chair and said, "I think first we had better take up your reports." Then he pointed at Jay who began: "Well, when Albert and I asked father for a piece of the garden for our very own to work exactly as we pleased this summer, with no questions asked or answered, he laughed. He said that Albert wouldn't stick to it a day."

"I will, just the same," shouted out Albert quite red in the face.

"Just keep out of this, Savage, until I get through."

The man laid a restraining hand on Albert's arm and Jay continued: "But I begged father, and told him we'd always worked for him, and he might let us try for ourselves. Besides, I told him we'd not shirk his work. So finally he said we each could have the ten hills in the corn patch for the experiment, just as you wished. And then—"

"And then," broke in Albert, unable longer to contain himself, "what do you think he gave us? It's just no use trying, for he gave us an old piece of land below the barn. It's a regular old swamp; why, water stands there the whole spring long, and it takes half the summer to dry it out. Then it gets hard as a brick. Now what is the use of trying on that?"

"We'll take it just the same, and so we told father," continued Jay. "We have just got to make that old land do something."

The Chief nodded and pointed to George for his report. "Same here," began George. "My father wouldn't listen at all at the first; then he said I might have the hills of corn. He threw in also an old side slope which he thinks is too poor for any use." George sank back in his chair in a quite dejected manner.

It was now Jack's turn. "You see," he began at his lightning rate, "we haven't much land anyway, seeing as we live in the village. I can have the backyard, such as it is, but that's precious little use. It's never been used for a garden, and it's full of rock. One of our neighbours says I may have a piece of her corn patch for my corn, if I'll take care of hers, too. Of course I took her offer. Just had to."

Peter took his turn last, as usual. "We have no place at our home; too much lawn, and mother will not have it cut up. Grandfather said I might have any place I wanted in his garden if I'd really care for it myself."

"My!" said Albert, "what a snap! Your grandfather has the best garden land around here. You'll win The Chief's five dollars for the best garden; you just can't help yourself. I'd like to punch you, Peter, for having such luck."

Peter smiled a little more broadly. "Fin sorry the garden's so good, but I can't help it."

"It's all right, Peter," began The Chief; "after all, boys, I believe we are not having such bad luck. Cheer up! We are going to surprise those fathers of yours, and have a good time out of it, too. Jay and Albert have a big problem of draining; George has simply got to put that sandy slope in shape; it looks as if Jack would have to fill in for his garden; and Peter—well, some of you may beat Peter yet."

At this last Peter smiled, Jack skilfully tumbled him off his chair and Albert gave a war whoop. The Chief called his meeting to order again.

"And now, boys, I shall see you each one separately about your garden problems. Remember, not a word at home, for we are going to surprise the people. And at our next regular meeting, and at all others this winter we shall have reports on the manner in which you are going to get at your work and the way in which you will beat conditions. In this way we can keep track of each other's work. We must make our plans, too, on paper, which will help out. We have catalogues to write for, garden stakes to make, and no end of things will come up. But first you boys ought to understand a bit more than you do about the soil. It is a storehouse of good things. Knowledge of the soil is a key to this storehouse.

"We can roughly divide the soil into three classes and call these sand, clay, and humus. The ideal soil has all three of these elements in it. Sandy soil is made up, as the name itself really tells, of broken up rock masses. One can tell this sort of soil by its lightness and the ease with which a mass of it drops apart. By the word lightness one does not mean colour or weight, but looseness. A clay soil may be told by its stickiness; its power to form lumps or masses; its tendency to crack and bake under the hot sun. Such a soil is called heavy. Humus soil is made up largely of decayed animal and vegetable matter. Its presence is told by a dark, rich colour.

"In trying to improve the soil we are dealing with, we have first to think of its physical, and second, its chemical condition.

"The great needs of the soil are air and water. Just think of all soils as made up of many particles; let us say like a lot of marbles, one placed upon another. Each given mass of particles has a given air space between every particle. Again, if a marble is dipped in water a film of water remains on it a short time. Let us think of the particles as always having a film of water on them. Then, as roots and root-hairs of plants strike down among these they find the two necessities, air and water.

"Now sand is very loose and so lets the water drain down through it too rapidly. How shall we improve a sandy soil? Just add something to bind the loose sand particles together. Humus is good for this binding purpose.

"Clay absorbs much water. Then its particles squeeze tightly together and so air is shut out. Add sand to clay soils, to lighten them.

"Humus soil is very rich in nitrogen. This brings us to the chemical side of soils. There are many chemicals in soils which act as foods to plants, but only three are the essentials. If these cannot be unlocked from the soil, or are lacking, they must be supplied. These plant foods are nitrogen, potash, and phosphorus.

"The chief source of nitrogen is manure; of potash, nitrate or sulphate of potash, and wood ashes; of phosphorus, bone ash or phosphates. How can you tell when one of these is lacking? Well, first it is well to know what each one does for a plant. Nitrogen makes fine, green, sturdy growth of leaf and stalk; phosphorus helps blossoming plants; while potash makes plump fruit. If foliage looks sickly then nitrogen is needed. If one wishes a good growth of leaves, as in lettuce, nitrogen is needed. If the fruit is small and poor, supply potash; while if the flower and stalk need better growth, add phosphorus.

"Be careful in adding nitrogen. Nitrogen is the plant food which is most easily lost out of the soil. Study the soil you are dealing with, supply if possible what it lacks, and try to unlock to your seedlings the goodness already in the garden soil."

The interesting talk ended.

"Couldn't we meet oftener than just Saturdays?" questioned Jack.

"We'll see; it all depends upon how much work there is. Possibly we may have to meet Fridays, too, later on, for you have no lessons that night. Anything more, boys, before the popcorn?"

"I'd like to know," asked Peter, "if my cousin Philip, who comes from the city to grandfather's to spend almost every Saturday and Sunday, may join us too. He wants to fix up his city backyard and doesn't know how to begin."

"Bring him along next Saturday. We'll be glad to have him, shan't we, boys?"

"Don't know," blurted out Albert, "it's our club."

"Keep quiet, Albert. Let him in as long as he behaves and works. Isn't that right?" asked Jay.

"Yes," answered George and Jack.

"Then, boys, we'll have Jack's report next week, as his problem is not so difficult. If Jay and Albert drop in some day from school they shall have a book which will help them, and George needs one, too."

At this point Albert dropped off his seat in mock alarm murmuring as he fell, "Worse, much worse than school!" George dropped a heavy book on top of him to add to his comfort.

The Chief went on as if nothing had happened. "Jack and Peter, shell and pop the corn, George and Jay, crack the nuts. And you, Albert, run to the cellar for the apples. Get good ones, young man."

"Why," questioned Albert, as he picked himself up, "why must poor Albert always do the hard work, while the other fellows stay by the warm fire?" No one answered him and he slowly marched off.

Soon the corn was popped, the nuts cracked and the big red apples on deck. And then it was a quiet room save for the snapping of a shell from a half-cracked nut, and the munching of the firm apples as the boys ate. The firelight played softly over the old room bringing out strongly the big oak table, the group of boys, the silent man, throwing far back into the shadows the old rush-bottomed chairs, the short-legged rockers and the pieces of furniture at all distant from the fire.

The clock struck nine. The boys reluctantly got up from the floor and struggled into their coats. Jay unbarred the door. The man held the light high above his head sending a stream of light after them, George astride his old farm horse ready for his three-mile ride, Jay and Albert trudging after him, and Jack and Peter hand in hand on a run toward the village.

"Good-night!" they shouted back at the man, "We'll be on time next Saturday night, seven sharp. Good-night!"

II

THE BOYS' GARDEN DIFFICULTIES

"It's fine to see you back, Myron," began The Chief, looking at a big, good looking fellow, who had dropped quietly into his place by George. "Are you here for all the time, now?"

"Yes," replied Myron, "I got tired of the town and am glad enough to be back again."

"We are just as glad as you are. Is Philip here for Sunday?"

Philip wriggled happily beside Peter and said nothing. For Peter had impressed Philip with the fact that he must keep quiet for it wasn't very much his club anyway.

"There is business up for discussion, and two applications," began The Chief.

"Applications!" broke in Albert. "May we have those first?"

"If you can't keep still," retorted Jay, "you'll get applications all right, but of quite another sort."

The Chief passed two letters to George. George stood up, swallowed hard, for he was a bashful lad, and began. "'Will the Junior Garden Club give suggestions and practical help for the improvement of the Oldfield Centre School Grounds?' Signed 'The Teacher.'"

"Crickey!" said Albert. "That's white in her! Expert advice! I guess we will!"

"What shall we do about this?" asked The Chief.

"We just ought to do it, I think," began Jack. "There surely is no other public-spirited club in this place."

"Just so," murmured Peter.

"We ought now to have a secretary for the club, and a chairman, too. I believe to-night is the time to vote for these officers," suggested The Chief. "Suppose Jack and Philip tear up slips of paper and pass them. Then Myron and George collect, and count the ballots. We should vote for chairman first."

"What does the chairman have to do?" asked Myron.

"A chairman always calls a meeting to order and presides," answered the man.

For a few minutes they were all very busy with paper and pencil. The results were given by Myron.

"Jay has all the votes for chairman. Albert has four for secretary, and so I suppose we'll have to have him."

Albert, nothing daunted, said, "I guess you will, but I write like a hen."

"That's right, you do," chimed in Jack to Albert's apparent annoyance.

"Now, Jack, call your meeting to order and let's have these matters voted on."

"Come to order all of you. What shall we do about this school-ground business?"

"I vote," began Myron.

"Stand on your feet," advised Jay.

"I vote—"

"No, Myron," corrected The Chief, "move—not vote."

"I move, then, that we fix up those grounds."

"Who seconds this?" and Jay looked hard at George.

"I will," he responded.

"I'm not sure, now," appealed Jay to The Chief, "I'm not sure just how to go on."

"It's this way—it has been moved and seconded that this request be granted. All in favour say 'aye'; all contrary minded 'no'. It is a vote." Jay repeated this and the boys voted, Albert, as usual, voting "no," just for fun.

"Now, if George will read the second letter— —-"

"I should think," Jack half questioned, "that the secretary should read things, now we have a secretary."

"So he should, hand those papers over, George."

George, delighted, gave place to Albert, who stumblingly read. "We girls wish to garden, too. We'd like to join your club. We can do just as good work as boys. Will you take us in?"

"Not much!" went on Alfred just as if he were still reading, "Girls in our club, no siree, girls never!"

"Girls might do something," began Myron.

"But," Jack broke in, "they'd after all spoil a boys' club. Why, it wouldn't be a boys' club then."

"They might have one of their own," suggested Peter.

"And do different things," continued Philip.

"I really don't see," Jay went on, "quite how we could have them. But, I suppose, they might meet with The Chief and we could help them sometimes."

"No," said Albert, "we don't wish to get mixed up with that sort of thing. They'd run the whole club in the end."

"That's right," agreed George.

This was put to vote properly and the girls were barred out.

"You must write them a letter, Albert," concluded Jay.

"Write a letter! A letter to those girls! Never!"

"Yes you will; you are the secretary and you have to, understand," said Jay. Poor Albert, not long before so proud of his office, looked as if he'd rather be whipped than be the secretary.

"The real business to-night is to hear reports so we can do a little experimenting and testing next week," suggested the man.

"All right, we'll have George's first."

"I shall, of course," began George, "plant my corn, Country Gentleman, in with father's. We have plenty of seed corn, so I shall not have to buy any. As far as my old slope goes I have to pick all the stone off. Then I am not sure just how to drain it, for the rains from another slope above wash it all the spring and summer. I shall then put some barnyard manure on and plant it all to corn. Of course, I must plough and harrow it, too."

"Now," said The Chief, "I guess we'd better stop right here and have a little talk, for George has brought up some problems for discussion. In the first place—let us consider the draining. All George has to consider is that he has to conduct or lead the water off his land."

"But," said George, "that is what seems to me difficult."

"Have you noticed how water takes definite courses down hills? That ought to give you some help."

"I see," cried Jack, "George could make gutters for the rain to travel along and so lead the water off his garden."

"Exactly, Jack has the idea. It is really a bit of engineering. Suppose George finds the highest point, the greatest slope, of his land. From this point a gutter or furrow should be dug so that the water is made to flow off and away from his land."

"How deep shall I dig the gutter?"

"Dig it about three feet deep and fill stones right into this gutter. Two feet of stone in the gutter is about right. Water falling on a stone mass drains off properly. It would sink into an earth mass. Bring a little sketch of this with you next week, George, showing where you are going to dig the drain. Now boys, how much fertilizer do you think ought to go on this poor land of George's?"

"I was going to put on two inches," said George.

"I should think he ought to put at least four inches on," half questioned Myron.

"I'll say eight," began Philip. The boys shouted at this.

"Philip," went on the man after the laughter ceased, "is very nearly right. If George wishes to get anything from this old land at once, he must fertilize it heavily. If your father can spare a foot of fertilizer put it on." The boys all whistled.

"Now about the corn! Did you know, George, that corn is a most exhaustive crop?"

"I don't even know what you mean."

"I do," said Jack, "he means corn tires the soil."

"Just so," continued The Chief, "the soil supplies food to the plant. Some crops use up more of the soil's goodness than others. Corn is one of these. Now, George, what do you think about planting a crop that works the soil very hard, especially when the soil you are dealing with is rather poor?"

"It wouldn't be the best thing, I should say. Will you suggest good things to plant?"

"Well, potatoes, tomatoes and cabbage demand less from the soil."

"Then I choose cabbage, I'll plant that entire old slope to cabbage."

"Now, Jack, suppose you talk."

"I have decided to build a coldframe, so I can get a little earlier start with my plants; I suppose I should have begun this frame last fall. I know this—that I have to dig out my whole garden spot and fill it in. So I thought I could get a start with the coldframe while I was working at filling in. I have decided to plant lettuce, radish, beets, tomatoes, peppers and some flowers. I think I shall plant asters, stock and sunflowers."

"Why sunflowers?" asked Philip.

"I want the seed for my squirrel's feed next winter. Then, too, I think sunflowers make a pretty nice background for a garden."

"If you wish to drop in to see me before the next week's meeting we'll have plans for the coldframe worked out to explain to the boys then. You measure the space where you are going to put the frame and ask your father about the lumber. As lumber is your father's business, I should almost think you could get us some soft wood, say white-wood, for our stakes and markers," suggested The Chief.

"Of course, I can," promised Jack.

"Now Peter," said Jay.

"My garden is to be just potatoes."

"Peter and potatoes!" jeered Albert. "Oh, Peter!"

"I don't care, I'm for potatoes and profit."

"Peter always does make money. So I suppose his potatoes will turn into money, too," volunteered Philip.

"My stunt," said Myron next, "is to be strawberries. I want to raise strawberries. Mr. Marsh, on the Longmeadow Farm, has offered to give me some plants. I'll do the corn stunt; aren't you going to, Pete?"

"Of course, that was understood, and Philip will have his corn at grandfather's too, for a city backyard is no place for corn."

"Now, Albert, you may talk for both of us," said Jay.

"Our land has to be drained, but it is not exactly the same proposition that George has. Water stands on our land. We had thought of putting a drain pipe in. It seems as if there should be an easier way, but we don't know one," Albert stopped and looked at The Chief, who leaned back in his chair and thought a minute.

"I guess, boys, we had better stop and talk over the matter of drainage. There are three kinds of drains, namely: the open drain, the blind drain, and the tile drain. Each one has worked out of the other. The simplest sort and the one man first used is the open ditch. A piece of land was covered with water. A ditch was dug through the land at the place or places where water was standing. Usually a little stone is thrown into the bottom to help drain the water off.

"Such a drain put out of use quite a bit of land. So partly because of this a second sort of drain was worked out. A good body of stone was put into the drain, then earth filled in over this. Water percolating down through the soil followed along these drainage courses. Formerly it settled in spots and made boggy land. Finally a more systematic sort of drain developed from this last one. Instead of a body of stone, a drain tile was placed on the bottom of the trench.

"Straight off you boys can see which one of these three represents the best all around drain. Out in the country or where there is no need to think of utilizing every bit of land, the open drain is often seen. But where every bit of land must be used, the open drain is out of the question.

"All drains come under the head of one of these three types. After all, boys, since you can put in the tile drain would it not be wiser to do so?"

"Surely," answered Albert. "But I should think soil which has been under water for some time, as this has, would be a bit poor."

"In case you find the soil is sour, as it may be, you can sweeten it up. There is a certain farm sweetener in lime," added The Chief.

"We shall plant on our land onions, peas, and tomatoes." continued Albert. "We believe that the soil is going to be especially good for onions."

"I guess I shall have to break in again right here. Onions need a fine, rich, deep soil. To be sure moist soil is good for certain varieties of onions. That is why, I imagine, you thought your soil good. You must get this soil into better garden condition before you devote it to a crop like onions. Try a general vegetable garden this season. Work out the crop value of the soil.

"Philip, do you know what you are going to do?"

"I know that I have everything to do. I thought perhaps I should do something like this. We want that old backyard to be really pretty. The yard is a long narrow strip of land just like most city backyards. I thought I'd make a walk straight through it. I want a little fish pond at the end. I thought I'd lay out a few flower beds with paths in between them. Mother says she will buy me a few shrubs."

"I say, Chief, don't you think some of us might go up to the city and help Philip make the cement pond?" asked Albert eagerly.

"We might," murmured Jay, "if we get invited."

"Boys, it's late. We know a little of what our stunts are to be. Next week each of you bring about fifty seeds of each kind you intend to plant. Be able to tell just how these seeds should be planted. Also have the dimensions of your plots. Jack will bring some soft wood along, too. And Philip, find out, if possible just how much money you can have for shrubs. Now on with your coats! Out of my house in two seconds!"

"No food to-night is a sad blow, Chief," said Albert pretending to weep as he opened the outside door.

"This blow is sadder," replied Jack, playfully shoving Albert clean out of the door.

III

THE GIRLS' SECRET

A very timid little knock roused The Chief from his study of Jack's coldframe plans. The outer door gently opened and three little girls entered and advanced to where the man sat. One, the smallest of the three, was thrust forward as spokesman. Gathering herself together she began with a rush. She thrust a letter into The Chief's hands.

"This is the boys' horrid letter. We don't care particularly about belonging to a boys' club. We wouldn't now, any way. But we'd like to show those boys a thing or two and we thought perhaps you would help us. Will you?"

"Sit down, and we will work out a little plot together. But first tell me your names. I like to know the names of people with whom I plot."

The girls came close to the man. The spokesman did the introducing. "I am Delia, Peter's sister, and just as smart as he is. This," pointing to a quiet, pleasant-faced girl, "is Ethel. And the other is Jack's sister, Elizabeth."

"How many more girls belong to this company?"

"They are all outside waiting, I'll call them in if you say so. They are behind the lilac bushes. You see we were afraid some of the boys might come to see you, so we hid. For we don't wish them to know about this at all. I'll call the girls in now." So Delia ran to the door, held it wide open, and called "Come girls, he wants to meet you!"

"Come right in, girls. This one," pointing to a girl with light hair and bright eyes, "is Eloise. Her father keeps the Inn. And this is Josephine, who has no yard at all; and Helena who has plenty of ground; and this," with a grand flourish, "this is the judge's daughter, Katharine."

"I hate," said Katharine, "always being labelled; I think it's pretty hard on a girl to be tagged this way."

"If you'll sit down," began The Chief—"although there are not chairs enough—we'll get right down to business."

And then how they talked! Closer and closer they drew up to The Chief until the eight heads were so close together they seemed almost one huge head. Finally they all shouted with laughter.

"Not a word outside, mind you, not one word. Prove that girls can keep a secret."

"We solemnly promise," said Katharine for the others.

"Look," cried Elizabeth, "there comes Jack; what shall we do?"

"Out this way," quietly replied the man, almost sweeping seven happy little girls out of the door. "Now, cut and run." And off they scampered over the fields.

IV

GARDEN EXPERIMENTS PERFORMED INDOORS

"The meeting is called to order," began Jay. "To-night, so The Chief says, each fellow has some special thing to talk about. Albert will have an accident with that bottle unless he begins right off, so tune up, Savage."

"This bottle is full of vinegar. I might have brought a lemon or anything else acid. This blue paper is called litmus paper. I got it at the drug store for ten cents. Just look right here, and you will see magic worked. I shall put some vinegar on this piece of paper. See!"

"Turned red as quick as scat!" said Jack.

"Litmus will always turn red when any acid gets on it. I've tried several acids at home. It works every time," went on Albert as if no one else had spoken.

"I cannot see what this has to do with gardens." began Philip.

"Now you keep quiet until I finish. Haven't you fellows heard your fathers talk about sour ground? Well, that means acid soil."

"Why, we have a piece of ground, where sorrel grows thick; father says that is sour," added George.

"Just a minute, Albert," broke in The Chief; "that is one way, George, that farmers tell a sour bit of land. Weeds grow thickly over such ground, but as George has said, sorrel is likely to predominate. Go on, Albert."

"Any soil may be tested with this litmus. The Chief calls this the scientific way of going at it. I was able to get a little soil from our future garden plot, and I'll find out right now if it's acid." Albert opened a small box which was full of soil that looked quite clayey. He wet a piece of litmus and buried it in the soil.

"We'll have to leave this a few minutes, and I'll finish what I have to say. If soil is very acid it has to be changed back again."

"Back again to what?" asked Jack.

"Why, back again, so it isn't acid," Albert continued, decidedly confused.

"I'll help you a bit," and The Chief came to the rescue, "Get that big bottle over there, Albert." Albert brought the bottle. In it was a liquid clear as water.

"Taste it, Peter," and The Chief handed Peter a little in a small glass.

"Why it doesn't taste like much of anything; sort of flat."

"That's it exactly, Peter. It certainly is not an acid, is it?" Peter shook his head. "It is lime water and does not belong in the acid class, but to one which is exactly opposite to the acids, the alkalies. Soils ought to be neither acid nor alkaline, but neutral, as it is called. An alkali will help make neutral an acid. If the soil is acid it is bad for your crop. Put a little lime water on the litmus which the acid has turned red."

Albert did this and the boys watched interestedly to see the effect. "Back again," sang out Jack as the red litmus changed to blue.

"Now from this you see a way to overcome the acid conditions of Albert's piece of land, if it proves to be acid."

"I see," said Jack, "lime it."

"Exactly! Now see, Albert, if the paper has changed colour."

"My, I should say it had!" and Albert held up the piece of litmus paper, now quite red from its contact with the soil.

"Well, Albert, it is pretty plain to see what you have to do. Did you find out the amount of lime to use?"

"In the book I read it said for clay soils 400-2000 pounds per acre."

"I should say," said The Chief, "for that special piece of land use about 20 bushels to the acre."

"How many pounds of lime," asked Jack, "to the bushel?"

"I can answer," grandly went on Albert, "there are 70 pounds to the bushel. So that makes 1400 pounds."

"Quite a proposition!" said Jay.

"Yes, but your land is only a half acre and so that changes matters a little. How much is lime a bushel, Jack? Ask your father, will you?"

"I think," said The Chief, "that we'll have to lay a drain pipe through your land. Anyway I shall come around in early spring and have a look at it."

"Now Peter, we'll hear from you," Jay announced.

"My work was to find out how long it took different kinds of seeds to germinate, that is sprout. I took a dozen each of different seeds, put blotters in dishes, wet the blotters, and placed the seeds on these. I kept them in a warm place in the dining room. I have made each of you fellows a copy of the table."

PETER'S GERMINATING TABLE

Beans	5-10 days.
Beets	7-10 days.
Cabbage	5-10 days.
Carrot	12-18 days.
Celery	10-20 days.
Lettuce	6-8 days.

Onion	7-10 days.
Peas	6-10 days.
Pepper	9-14 days.
Radish	3-6 days.
Tomato	6-12 days.
Turnip	4-8 days.

"I'd like to know what use a germinating table is, anyway?" asked Albert scornfully.

"Well," Peter replied thoughtfully, "it gives you an exact knowledge of the time to expect your seedlings to poke up. Now suppose radish came up in four days. The carrot you had planted didn't come up and after twelve days, discouraged, you plant more seed. Now two days later the first carrots you sowed begin to appear. If you had known that it took carrots from 12 to 18 days to germinate you'd not have made the mistake of planting again so soon. I think of another reason," went on Peter warming up to his subject. "Suppose you planted beet seed. You waited ten days; nothing happened; you wait two more and still no seedling appears; something is surely wrong and you plant over again."

"What could be wrong," asked Philip.

"The seed might be poor," replied Peter. "George has been testing seed," said Jay, "and he might tell us about it now, couldn't he, Chief? It seems to come in here." The Chief nodded.

"I have been finding out whether certain seeds which I happen to have on hand are worth planting or not. If any of you fellows have seed and wish to find this out, you can easily enough. So you can be sure whether old seed is worth planting. Now it happens that father had some of his last year's corn and some from four years ago. So I took 100 seeds of each. If you test small seed like lettuce, The Chief says 50 seeds will do. These I put on blotters just as Pete did his. Of course, I kept them separate. From last year's seed 90 seeds sprouted out of the 100, or 9/10 of them. And that equals 90 per cent. If all seed was 90 per cent. good it would be all right to use, I think. Now when I looked at the four-year-old seed, what do you think? Only five seeds had started. That makes only 1/20, or 5 per cent. Of course, no one would care to use seed where only 5 per cent. of the seed sprouted."

"Is there any real percentage of germination that seeds should have?" Jack asked eagerly.

"Yes," replied The Chief, "although value as you see from George's experiment is lost by age. The real standard germination value for corn is 87 per cent., for beans 90, for turnips 90, for peas 93, etc. You can see that the per cents. for these vegetables run high. So do not use seeds when the per cent. has dropped too low.

"Has George found out the time when other seeds lose value?" asked Peter. "I did not work this table out because I did not have the old seed to work with," replied George, "but The Chief gave me a book to look it up in. I have printed on our press the table. So you fellows may each have a copy." George handed the sheets around the table.

It happened that The Chief had a little old printing press that he had presented to the Club. Club real estate, Albert called it.

GEORGE'S TABLE

AGE OF SEEDS FOR PLANTING PURPOSES

2-3 years.	**3-4 years.**	**5-6 years.**	**8-10 years.**
Corn	Beet	Pea (5-6)	Cucumber
Celery	Lettuce	Radish (4-5)	Melon
Carrot	Turnip (3-6)		Squash
Bean			Pumpkin
Tomato			
Pepper			
Onion			
Parsley			

"Now, George," Albert begged, "give us a table of germinating per cents."

"Not much, each fellow can work out the value of his own old seeds and see if they are worth using."

"I think George is right," began The Chief after the laugh at Albert's expense ceased. "Perhaps you'd like to try the effect of depth of planting on corn. Here are some boxes of earth. George, you plant six kernels of corn one inch deep and mark the box with your name and the depth on it, Peter, plant the next box with six kernels at two inches. Albert, try three inches, and Jack, four inches. It will be your business, Myron, to drop in here each half day and note the first appearance of corn in the different boxes."

The result of this experiment, which took about two weeks in all, was as follows:

DEPTH OF PLANTING	TIME TO COME UP
1 in.	8½ days
2 in.	10 days
3 in.	12 days
4 in.	13½ days

This experiment showed the boys that seeds too deeply planted are hindered in progress.

"Myron, you may take the floor now," signalled Jay.

"I have worked out and printed for you the amount of seed necessary to plant a certain space. I have printed my table just as George did. 'H' stands for hills and 'D' means drills."

"What is a drill?" asked Philip.

"Why a drill is a furrow. You can make a drill with a rake handle, or a hoe. We can show you better when we get outdoors, Philip," Myron answered quite condescendingly.

MYRON'S SEED-ESTIMATE TABLE

NAME	METHOD OF PLANTING HILLS OR DRILLS	QUANTITY OF SEED
Bean (Bush)	D	1 qt. for 100ft.
Beet	D	1 oz. for 50ft.
Cabbage	H	1 oz. for 2000 plants
Carrot	D	1 oz. for 100 ft.
Corn	H	1 qt. for 100 hills.
Lettuce	D	1 oz. for 120 ft.
Musk melon	H	1 oz. for 60 hills.
Onion	D	1 oz. for 100 ft.
Parsley	D	1 oz. for 150 ft.
Pea	D	1 oz. for 100 ft.
Pepper	D	1 oz. for 2000 plants.
Potato	H	1 peck for 100 hills.
Pumpkin	H	1 oz. for 30 hills.
Radish	D	1 oz. for 100 ft.
Tomato	H	1 oz. for 1000 plants.
Turnip	D	1 oz. for 150 ft.

"This table is all right, I suppose," began Philip, "but if a fellow doesn't know quite how far apart to plant his cabbage, say, I can't see how this table helps much."

"I took it for granted," Myron answered, "that you fellows know a little about things. But if a person didn't know what you ask, Philip, I suppose this table isn't much good. Shall I call all the tables in, Chief?"

"Not at all, Myron, this is a good table so far as it goes. Next time each of you boys look up this matter. Perhaps you can work out a good scheme for such information."

"Now, Philip, we'd like to hear about your shrub money and then we'll have time to see Jack's coldframe plans, before club time is over," at which Jay settled back in his chair as if club work was a strain on a fellow after all.

"I may have one dollar to spend. I have decided to buy three shrubs. I shall plant one by itself; the two others together in a clump. I wanted forsythia, but I have finally decided on Japan snowball and Van Houtte's spirea."

"Why?" asked Albert.

"You see the forsythia shows up best against a dark background because of the bright yellow flowers. I have no good setting for such a shrub. Then, too, it blossoms so very early in the spring, in April you know, that it seemed to me, since I must plant this spring, I'd disturb less a later flowering shrub. I chose the Japan snowball because it's less liable to have lice than some others and because it looks well all by itself on the lawn. That spirea is a specially good variety of spirea because it does well almost anywhere, and also it is very showy and the foliage is handsome all summer long. Some shrubs look scrubby after awhile."

"Where did you get all this knowledge, Philip?" asked Albert, half enviously.

"I made it my business to know. I hunted up shrubs in a catalogue, then I called on a florist, and we had a shrub talk together."

"Now, I call that getting down to real work," Jay remarked. Philip looked happy and Peter nearly tilted his chair over in his pleasure for he evidently felt the city was making good.

"Now, Jack, bring on the coldframe."

"I have my drawing right here," began Jack, spreading it out on the table while the boys crowded about. "You look at the drawing as I explain. Myron and Jay have promised to help me make it. It will be a coldframe this year; next fall I shall change it into a hotbed."

"How?" broke in Albert.

"I shall dig out the soil from the coldframe. Then I shall put in two feet of manure and cover it with four inches of soil. This spring about all I can do is to mix into the soil some well-rotted manure. I guess I shall put in about three inches in all. I guess I can explain," continued Jack, delighted at this opportunity to air his newly acquired knowledge. "The Chief has talked this over with me. It all depends upon what you wish to use the frame for. I want to use mine to get an early start this spring, so I make the bed rich and depend on the sun's rays mostly for heat. This, then, is a coldframe. The sloping glass frame helps you see. But next winter I hope to really get results out of this frame, so I have to supply extra heat. The layer of manure underneath gives this. I then have a hotbed. If I just wish to keep plants along, ready to force next spring, then the sun's rays would be enough for that work without the layer of heat."

"I see, thank you, and why do you say layer of heat? I should call it a layer of manure."

"Because it is heat, isn't it? And anyway real gardeners call it that. We may as well use the right names; don't you think so, Chief?"

"Surely, Jack. It's our business to know right terms. Each line of work has its own language. Jack has done a good piece of work so far. We shall have most of our next meeting in the workshop. Jack, Myron and Jay are going to work on this frame. You other fellows will be able to make stakes and dibbers enough for the crowd."

"What is a dibber?" asked Albert.

"That is for you to look up. If you have any old rake or hoe handles bring them along for dibber making. Good-night, boys."

Off into the night they scampered—a jolly, sound lot of lads.

V

THE WORKSHOP END OF THE GARDEN

"Before we go to the workshop we might take up the methods of planting our vegetables. Then if any fellow has worked out a table, Peter, the star printer, may strike off copies for all of us," began Jay, after calling the meeting to order. "I'd be glad to hear from any of you fellows who have done anything on this matter."

All was quiet. Finally Myron arose and began to read from a paper covered with writing. "The carrot—common name of the *Daucus Carota*—a biennial, indigenous to Europe, believed by some botanists to have been derived from the common wild carrot."

"Where'd you copy that stuff? No table can be made from that! Imagine a fellow out planting carrots and reading before he sows: The carrot—a bi—bi what, biped, did you say, Myron?"

Albert chuckled away and Myron dropped into his seat saying angrily, "I tried hard, anyway. It took me a whole evening to copy just the carrot."

"I should think it might have. Has any fellow a really simple table?"

"I've worked at it," Peter replied modestly. "I think I have something here that will really be of use." At this Peter spread out on the big table a neat piece of work.

PETER'S OUTDOOR PLANTING TABLE

NAME	DEPTH TO PLANT	DISTANCE APART	
		SEEDS	FURROWS
Bean (Bush)	2 in.	12-20 in.	3 ft.
Beet	1½ in.	4-9 in.	12-15 in.
Cabbage	½ in.	20-24 in.	3 ft.
Corn	1½ in.	3 ft.	3-4 ft. (hills)
Lettuce	½ in.	6-8 in.	12-18 in.
Musk melon	1 in.	4-6 ft.	4-6 ft. (hills)
Onion	½ in.	4-12 in.	10-12 in.
Parsley	½ in.	6 in.	1 ft.
Pepper	½ in.	18 in.	2 ft.
Potato	5 in.	12-18 in.	24-36 in. (hills)
Pumpkin	1½ in.	8-10 ft.	8-10 ft. (hills)
Radish	½ in.	3 in.	6-8 in.
Tomato	½-1 in.	3 ft.	3 ft. (hills)
Turnip	½ in.	6 in.	12 in.

"That's all right," and The Chief laid a hand on Peter's shoulder and he smiled across at Myron.

"Each one of you boys ought to know how to make a working plan of his garden. I showed Jack how to make his coldframe plan. It is well done. Now gather about the table and I will make a plan of a supposed garden."

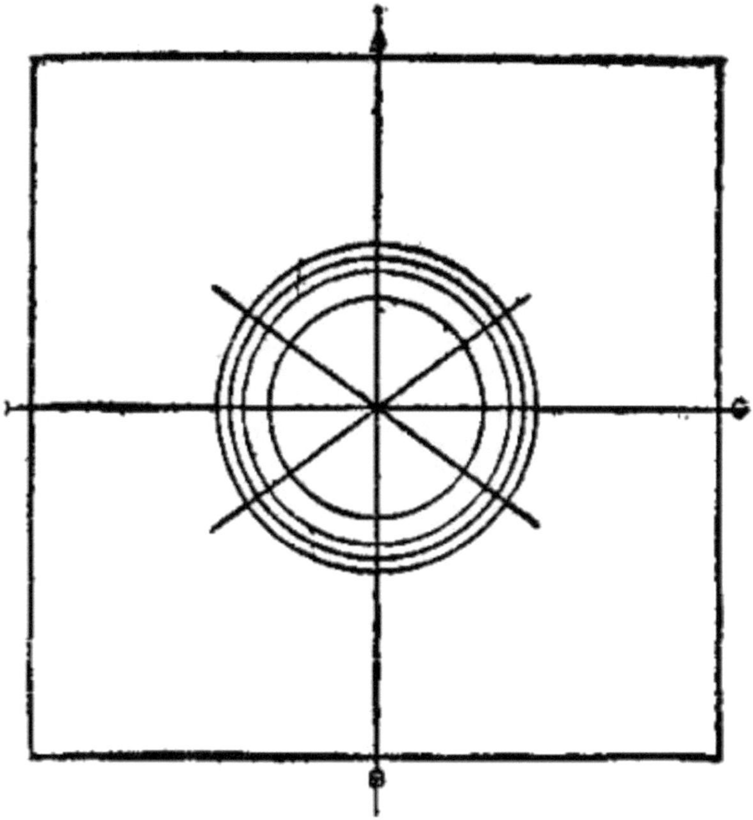

DRAWING I

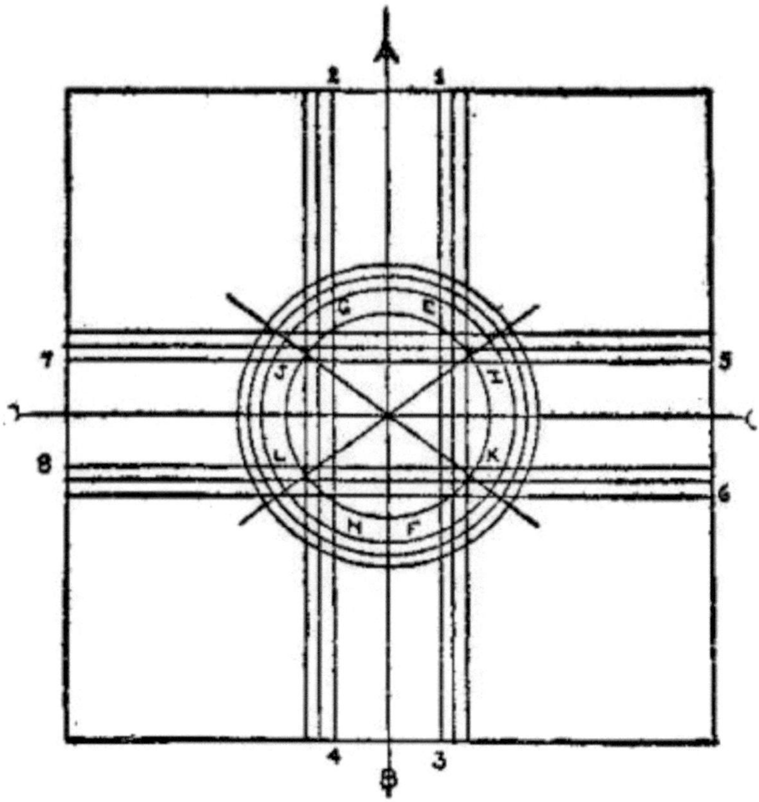

DRAWING II

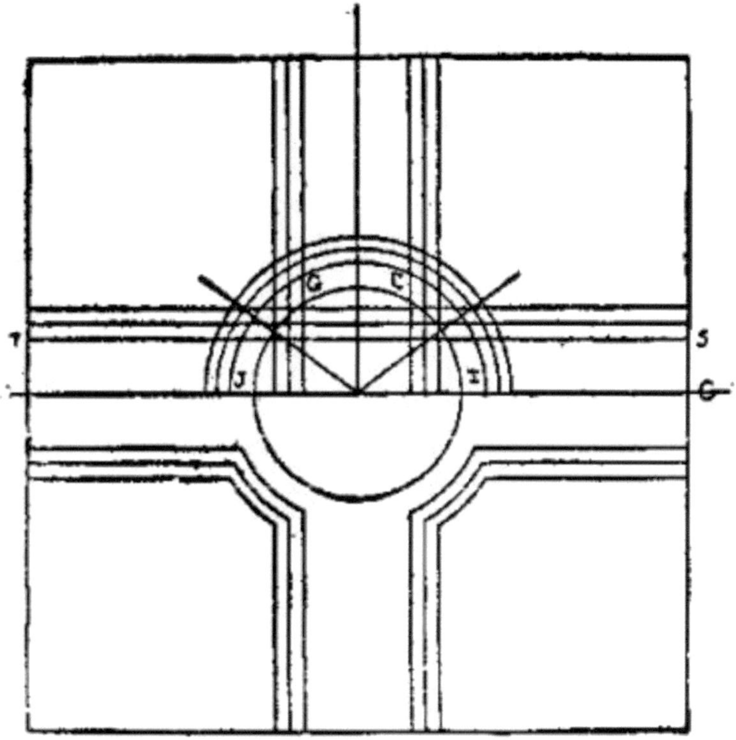

DRAWING III

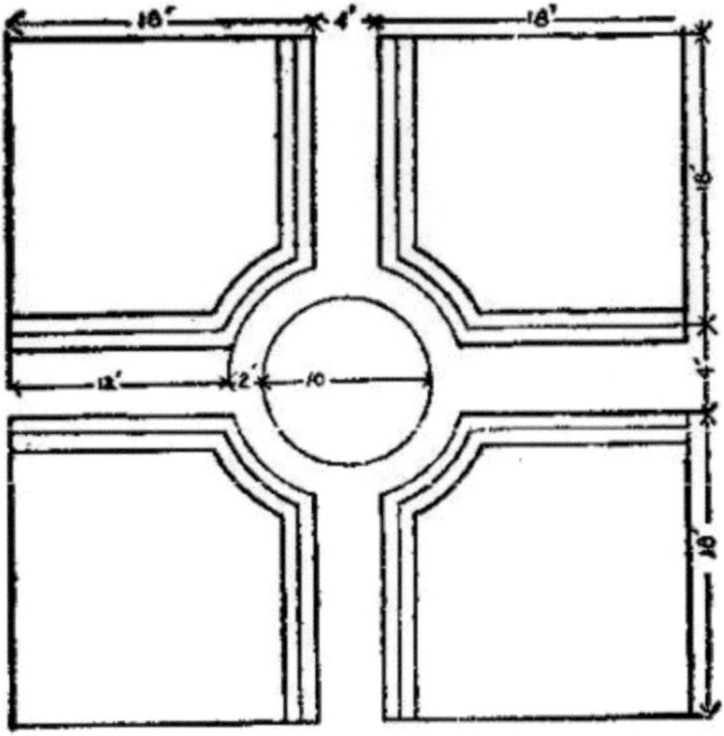

DRAWING IV

This very simple plan of a garden, used by The Chief, has in it the essentials for all your garden plan drawing. Follow each step as the boys did and you will be able to make a drawing of your own garden.

"I will lend you mine and you might make a drawing of that," craftily suggested Albert.

"No, young man, you are to make your own. Let us suppose for the sake of an easy problem that we claim our garden is to be on a square piece of land, forty feet by forty feet. In drawing to a scale, one takes a certain small measure to stand for a foot. If we take an inch to be a foot, then the entire forty-foot length would have to be forty inches. That is a pretty good large drawing. Let us take something smaller and say one-eighth of an inch equals a foot, thus ⅛ in. = 1 ft. So we shall have a length and a width of five inches.

"The first step in the actual drawing is to find the centre of your given piece of drawing paper. See, I just make short lines or portions of diagonals through the centre as shown right here in what I call Drawing I. Draw a vertical line through the centre extending to the top and the bottom of the paper. Now draw a horizontal line through the centre to the extreme left and right of the sheet. Now measure up from the centre on the vertical line the half width of the garden. If the centre is to stand for the centre of the garden, then the garden itself would extend up, down, and to the right and left of its centre, just 20 ft. or 2½ in. in a plan with scale ⅛ in. to 1 ft. So measure up from the centre along the vertical line just two and one-half inches and place a dot. Letter this dot A. Do this same thing down the vertical line and we have dot B. Also measure the same distance along the horizontal to left, calling the dot D and along the right calling the last dot C. Now draw a horizontal line 5 in. long through A with 2½ in. either side of the dot. This gives you one side of your garden or a 40-ft. length. Do a similar thing through dot B. Through C and D draw similar lines. We now have the outline of our garden of 40 ft. square. We have on our paper, though, a square 5 x 5 in.

"I have decided to have a circular bed in the centre of the garden which shall be 10 ft. in diameter. Therefore, the radius of the circle should be 5 ft. or ⅝ in. Get a pair of compasses for that, Jack. Now I shall swing the circle. But I wish a 2-ft. path all about this circular garden. If the path is 2 ft., then I must set my compasses on 2/8 in. more or now make the ⅝ in. into ⅞ in. Let us swing another circle with the same point as a centre.

"It strikes me that if I should lay my garden out into four squares, the combination of squares, central circles and straight main paths would look incongruous. So I shall cut the central points of the four square beds off by swinging circles. Have patience and you will see, for the general plan is in my mind just as it ought to be in the mind of any person who is to make a garden. Now swing another circle with a radius of 1 in., and still another the radius of which shall be 1⅛ in.

"Now we come to stage two of this working drawing (Drawing II). I wish a 4-ft. path going down to the centre bed from the points A, B, C and D. Place your ruler, Jay, on point A, for you may draw now. Measure to the right of A ¼ in. and to the left ¼ in., and place dots at these points. You have the width of your 4-ft. paths. Do this same thing at points B, C and D. Number these points 1, 2, 3, 4, 5, 6, 7 and 8. With very light lines connect points 1 and 3, 2 and 4, 5 and 7, 6 and 8. Where the line 1-3 cuts the second circle from the centre, letter the intersections E and F. The intersections of 2-4 mark G and H, of 5-7 I and J, and of 6-8 N and L. You now see the outline of these paths running through the garden. Let us border each path with two 1-ft. borders. So place the ruler at point 1 and mark off two ⅛ in. spaces by dots. Do the same at points 2, 3, 4, 5, 6, 7 and 8. Connect the opposite dots by light lines.

"Now let George take the third stage (Drawing III). Go right over the inside circle so as to make it stand out boldly. Strengthen line 1 to E, 2 to G, 3 to F, 4 to H, 5 to I, 6 to N, 7 to J and 8 to L. Now these circles should be strengthened and lines erased that interfere. That leaves curve EI, GJ, LH, and FN standing out clearly. You see in the drawing one-half the garden plan erased and all right.

"After Myron has erased every line (Drawing IV), you will see the garden plan in all its neatness. Place the measurements on the drawing. It looks well, does it not, boys?

"These are the steps. Any of you can work out your plan if you have one to work out."

"Now boys, for the shop! Myron, Jay and Jack are to work on the coldframe. Peter will have an evening's work printing this planting table. Albert will tell us the use of the dibber and make you one each from all these old handles."

Albert, assuming a grand oratorical manner, gave the boys the benefit of his search for knowledge. "A dibber is a pointed tool, usually a stick, used to make holes for planting seeds, bulbs, setting out plants and transplanting of seedlings."

Off they all trooped to a little workshop back of the man's home. Soon the boys were hard at work, sawing, whittling, and setting up type.

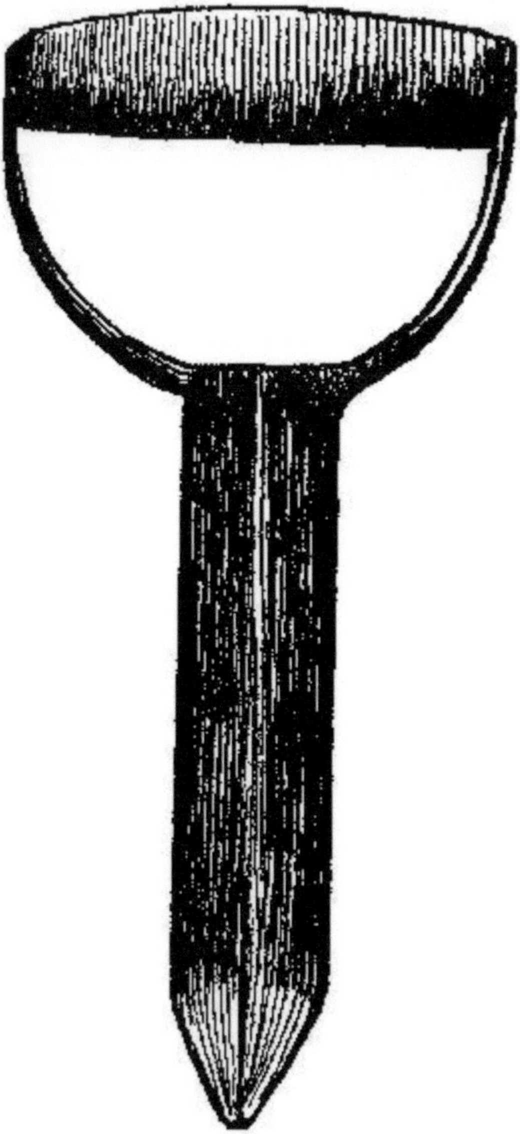

A sturdy dibber. This needs no directions for the making. The cut tells the story.

Here are directions for what the boys made.

DIRECTIONS FOR MAKING A COLDFRAME

Hemlock was the wood Jack used. The lumber for sides and ends is one inch thick while strips marked A and B are one and one-fourth inches thick. Cut out pieces 14 in. x 5 ft. 7 in., and 10 in. x 5 ft. 7 in., for the back and the front. Cut two pieces 14 x 36 in. and shape them according to drawing for the ends. Nail these four pieces together to form the frame. The sides should be nailed to the ends. Use ten-penny nails and drive them slantingly.

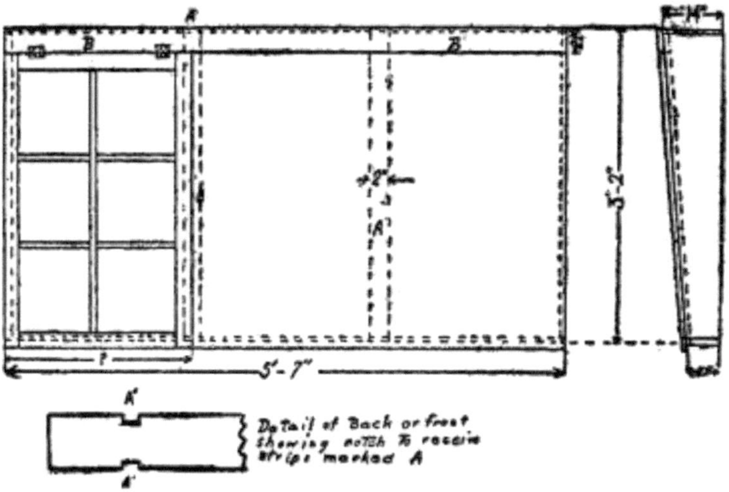

WORKING SKETCH OF HOTBED Jack's working drawing of his hotbed. This was to serve for a coldframe temporarily.

Saw out strips A two inches wide and as long as the slanting edge of the end of the frame. Be careful with this measurement not to measure the slanting edge of the *end* piece only, but to include with it the thickness of both front and back pieces. Saw out two more pieces two inches wide and as long as the frame is wide at the bottom. Make strip B 2½ in. x 5 ft. 7 in.

Lay out notches marked A by dividing top and bottom edges of front and back into three equal spaces. Cut notches to receive strips marked A. Nail strips A in place, also B. To make a neat piece of work the ends of strips A should be planed slightly slanting to make them exactly even, or "flush" with front and back boards. The real object of strips A is to keep the frame from bulging at the centre.

Jack had three common single window sashes, 22-in. x 3 ft. which made an excellent cover for the frame. These should be placed in position and fastened to strip B with two-inch butts. Notice the sashes project over the front so as to carry the water away from the frame. The sash should be fastened to the frame, putty side out.

GARDEN STAKE

The stake may be made of soft wood or hard. It is a good one to use in staking off the garden. It is entirely a piece of knife work. The dimensions are clearly given on the working plan. If the stake is made 12 inches instead of 14 inches, it may be used as a foot rule in measuring off furrows.

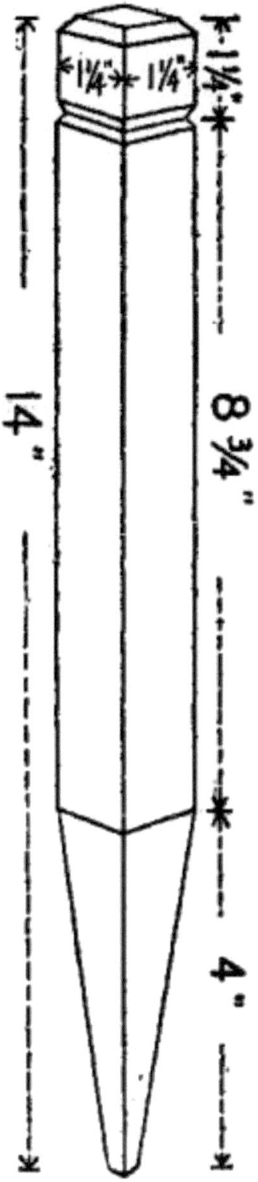
1¼"
1¼"
1¼"
14"
8¾"
4"

A heavy stake like this one, Jack used in marking off his drills. It is adapted to just such work.

THE BOY'S GARDEN REEL

A piece of wood (ash is suitable for this work), 11⅜ x 4½ x ½ in. is needed. Draw pencil lines lengthwise and widthwise through the centre of this piece. From the centre measure out one inch in both directions, placing dots. These give the central points for centre cut.

Measure from the four corners of the piece 3 inches along the length. Connect by line the opposite dots. This gives the line marked 4½ in. in the diagram. It shows the beginning of the cut to the centre line. One inch above these lines draw other lines straight across the wood. Find centre of these. Place a dot one-half inch on both sides of each from centre. This gives the one-inch end cuts. Cut this up to one-half inch of each corner. This makes a large substantial garden reel.

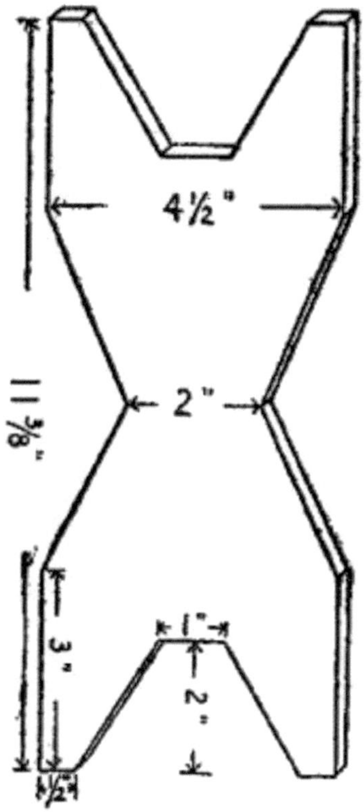

Made of hard wood this reel will last forever.

A PLANT STOOL OR TABOURET

The materials needed are four pieces 18 x 3 x ⅞ in. planed, for legs; one piece 14 x 14 x ⅞ in., planed, top; two pieces 8½ x 1¾ x ⅞ in., planed, lower braces; one piece 8½ x 8½ x ⅞ in., planed, upper brace. Use chestnut, white wood, white oak, mahogany, cherry or birch. You will need also 2-in. blued screws, round head, for fastenings.

This looks like a pretentious piece of woodwork for a lad to make. George did not think so. The construction is simple. Note the good lines.

To construct the stool make with the two $8\frac{1}{2}$ x $1\frac{3}{4}$ x $\frac{7}{8}$ in. pieces the lower braces, a lap joint. Find the mid-line of each piece by measuring $4\frac{1}{4}$ in. from the ends. From this line lay off two other lines parallel to it and at a distance of $\frac{7}{8}$ in. to the right and left. This makes a $1\frac{3}{4}$ in. square in the centre of each piece. Now transfer these lines down the edges of the lower brace pieces. Saw on the inside of the lines down one-half the thickness or saw and chisel down to one-half. It is necessary to saw on the inside of the lines or a loose joint will result. The joint must be exactly in the middle and all arms must be equal in length when completed. Brads or finishing nails should be used to hold the joint in place. This lower brace is 7 in. up from the floor or bottom of the stool. In the picture the screws, which hold the brace, show plainly.

Now lay off an octagon, with a diameter of 8½ in. on the 8½ x 8½ x ⅞ in. piece, sawing off the corner pieces so as to just fit the leg. Glue and screw this to the under sides of the top piece, placing the grain across that of the top wood. Warping is thus prevented. This brace acts as a support to which the upper ends of the legs are firmly screwed and glued. A 3/16 in. gimlet hole should be bored for each screw or the wood will split. The holes should not be deeper than 1½ in. if the screws are to hold firmly. Try drawing the screws across a cake of soap and see if they will not be applied more easily.

To be sure that the legs go on exactly rigid it would be well to draw lines diagonally through the centre of the under surface of the top piece. The legs are to be attached at right angles to these diagonals. After the legs are screwed to the upper and lower braces sandpaper the entire stool. Do this lengthwise to the grain, never across. Then stain and wax.

VI

WHAT THE GIRLS MADE WINTER EVENINGS

While the boys were making their pieces of garden apparatus the girls were at work also. They met with The Chief at Katharine's house and made a number of pieces of garden apparatus. The directions for making these are given so that other children may make some too.

DIRECTIONS FOR MAKING SEED ENVELOPES

Cut paper 7½ in. by 5¾ in.; place it the long way of the paper going from front to back of the desk or table at which you work.

Measure from the upper left corner down 1¾ in., and place point 1; 3¼ in. farther down place point 2. Measure from the upper right corner down 1¾ in. and place point 3; 3¼ in. farther down place point 4.

Measure from the upper left corner toward the right 1¼ in. and place point 5; 3¾ in. farther toward the right place point 6. Measure from the lower left corner toward the right 1¼ in., and place point 7; 3¾ in. farther toward the right place point 8.

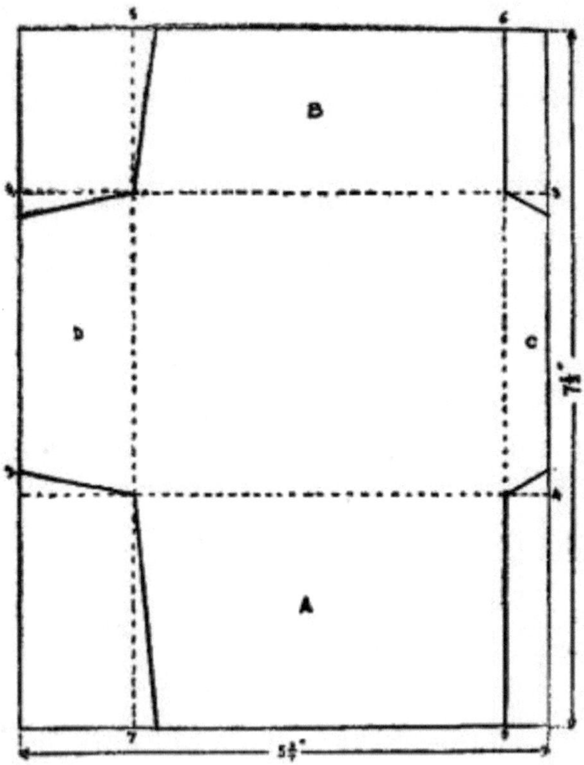

PATTERN FOR SEED ENVELOPE Katharine made this seed envelope of rather stiff paper So it was of real service

Draw dotted lines through 1 and 3, 2 and 4, 5 and 7, 6 and 8.

Measure ¼ in. toward the right from points 5 and 7 and place a dot. Draw full lines toward the left to the intersection of the dotted lines. Measure ¼ in. down from 1 and 3, and place dots. Draw full lines upward to the intersection of the dotted lines. Measure ¼ in. up from points 2 and 4, place dots, and draw full lines downward to the intersection of the dotted lines.

Draw a full line from points 6 and 8 to the intersection of dotted lines.

Cut on full lines.

Fold on dotted lines.

Fold A, B, and C, in this order, and paste, leaving D for flap to be pasted down when the envelope has been filled with seeds.

DIMENSIONS OF MARKERS

The right marker is 3½ in. long. The distance from head to central point of notch is ½ in. The distance between notches, or from the central point of one notch across the marker to the central point of the other, is ⅜ in. The width is ½ in. and the thickness ⅛ in.

The middle marker is 4½ in. long, ½ in. wide, and 1-1/16 in. thick. Allow about ⅝ in. for the pointing at the end.

The left marker is rather larger and stronger; it, too, may be pointed and not notched, so acting as a good pot-marker. Make it 5 ins. long, ⅞ in. wide, and 3/15 in. thick. The line between the notches measures ⅝ in. and is 1 in. from the top of the marker.

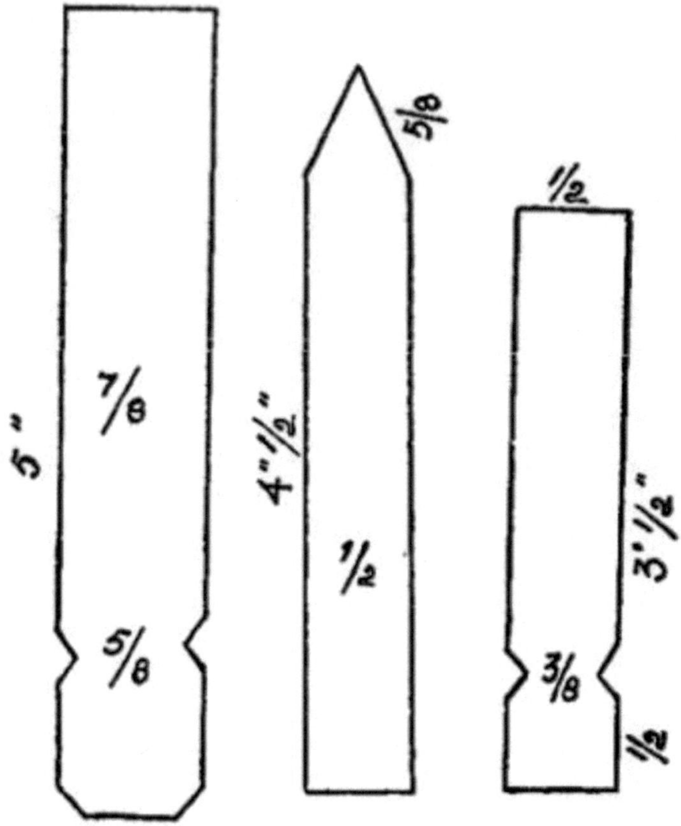

Such labels as these, made of thin wood, serve as plant labels as well as being useful in the outdoor garden.

A GARDEN SIEVE—MATERIALS

2 small boards 13 x 2½ x ½ in.
2 small boards 7 x 2½ x ½ in.
2 strips of wood 12 x ½ x ¼ in.
2 strips of wood 8 x ½ x ¼ in.
Fine wire netting 13 x 8 in.

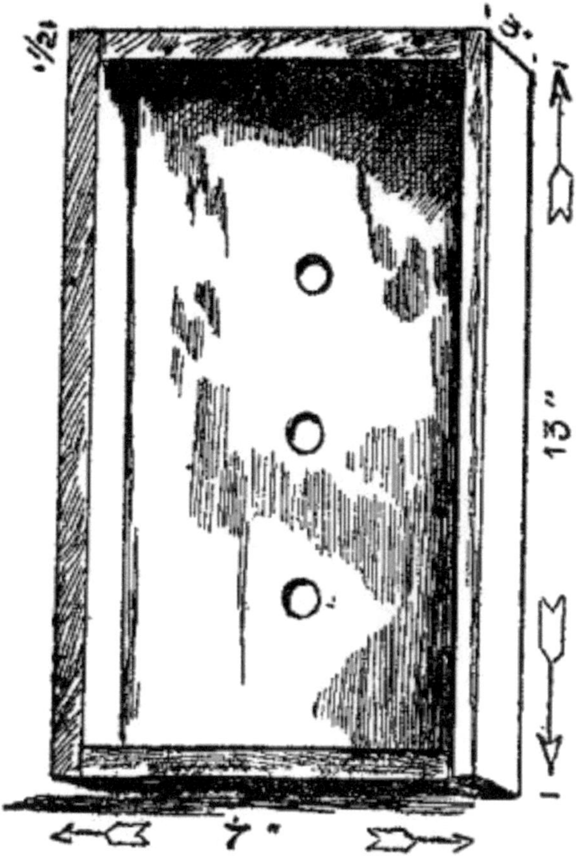
1/2
4
13"
7"

Josephine's box had too limited a drainage area

Make a flat like this one of Eloise's and so provide plenty of drainage space

Make the framework of a box without a lid, using the 13-inch pieces for the sides and 7-inch pieces for ends, putting the ends between the side pieces. Use the wire netting for the bottom of the box, nailing it on with the strips of wood. Paint the sieve with two coats of dark green paint.

A BULB FLAT

The dimensions of the box are the same as those for the sieve except for the depth, which is three inches instead of two and a half inches. Of course the bottom is wood with three drainage holes bored in it. A flat may be constructed without the drainage holes as shown in the cut. In this case make the bottom of small pieces of wood leaving an inch space between each piece. This is Eloise's kind of a bulb box.

A GATHERING BASKET FOR FLOWERS

The materials needed are:

8 spokes, 10 ins. long, of number 6 reed.
3 weavers of number 2 reed.
12 weavers of number 3 reed.
31 spokes, 20 in. long, of number 4 reed.

Directions.—Split four spokes of number 6 reed exactly in the centre, and slip the remaining four through the slits in the first group.

This is the basket made and used by Katharine. It is a gathering basket just right for fruit and short-stemmed flowers

Double a number 2 weaver and slip the loop over the upper vertical group and with the pairing weave go around each group four times. Next, separate the spokes in groups of two and continue the pairing weave until four more rows have been woven in. Then separate the spokes by ones and weave until the diameter is 4½ in.

Cut all off that remains of the number 2 weaver, and insert 3 weavers of number 3 reed. Continue with the triple weave to a diameter of 9 in.

Cut off the ends of the spokes and insert 31 spokes, 20 in. long, of number 4 reed; one on each side of the spokes, except the first; in this instance insert but one.

Use the side which has been next the weaver for the inside of the basket, letting rough ends come on the outside of the basket. Turn the spokes up, and hold in place with one row of quadruple, weave over three spokes and back of one, using the number 3 reed. With the same reed put in eleven rows of plain weave, over one spoke and under the next. Next, one row of quadruple and follow with seven rows of double weave, over two and under one, and finish with one row of quadruple weave.

For the first row of the border carry number 1 spoke back to number 2 spoke, or the next spoke at the right, and out; number 2 spoke back of number 3, and out. Continue once around the basket.

For the second row carry number 1 spoke over number 2 and 3, and down; number 2 over 3 and 4 and down, and so on around.

For the third row carry number 1 over number 2 and down; number 2 over number 3 and down. This may be continued until you have formed a roll over the entire edge.

If handles are desired, on each side of the basket insert a piece of number 9 reed for the foundation of these. The end of a number 3 weaver is woven in at the left of the foundation under the third row from the top of the basket, and the long end of the weaver is twisted around the foundation to the other side of the handle. Here it is pushed down inside the basket on one side of the handle and over again on the other side of the handle, three rows from the top, making a loop inside. The weaver is then laid close beside the first twist and follows it across to the opposite side. Now it goes in under the third row on the left of the handle and out on the right side. Each row of twisting must follow close beside the last. Six or seven rows will cover the foundation. The end is fastened off by bringing it inside the basket again where it is cut off.

This flower basket may be made without the handles. But they add much to it without being a great deal of extra work.

A SUNDIAL

Take two pieces of the wood you have chosen:

A, 6½ x 6½ x ¼ in. and B, 7 x 7½ x ¼ in.

Construction.—True up each piece to the given dimensions, and sandpaper carefully. Be careful to stroke the wood always with the grain—never across the fibres.

Dee's sundial kept fairly accurate time. It is a real ornament to the garden.

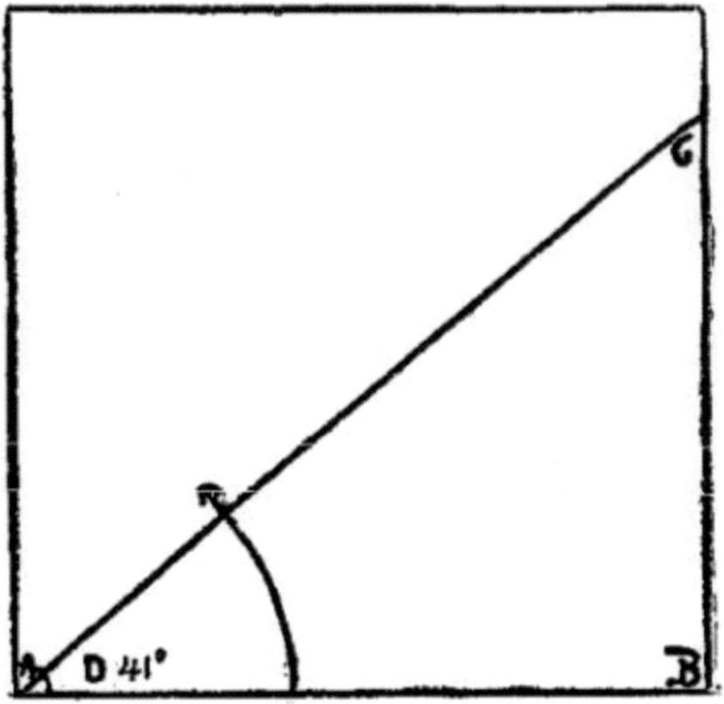

Gnomon pattern just one-half actual size

Next make a shadow-piece, or gnomon, as it is called. Get a thin piece of the same kind of wood as is used in piece A, and lay it out as follows: With the fibres running in the direction AB, beginning at point A construct an angle equal to the latitude of the place where the dial is to be used. For example, if the latitude of a town is 41 degrees construct the angle D 41 degrees, or if it is 42 degrees, let D be 42 degrees. Then cut from A to C, and sandpaper carefully. Take the wooden shadow-piece and fasten it to the centre of piece A. Fasten by two brads or small nails about ¾ inch or 1 inch long, or glue it. Place piece A over piece B so that a margin of ¼ in. will be left on all sides.

Place A so that the fibres will run at right angles to B to prevent the boards from warping. These two pieces may be fastened together by driving a brad in each corner, or gluing, or both.

POT-REST

Use almost any kind of wood, as white wood, cherry or white oak.

Two pieces of wood 8 x ½ x 1½ in. are needed for the cross pieces. These should be planed. There are needed also four little pieces as feet or pads. The dimensions of these should be 1½ x 1½ x ½ in.

To make this stand, draw a line across the two long pieces 4 inches from either end. Lay off two other lines parallel to this ⅝ inch to the right and left. Transfer these lines down the edges by the aid of the try square. Saw on the inside of these lines down one half the thickness, or ¼ inch. Chisel out for a half-lap joint.

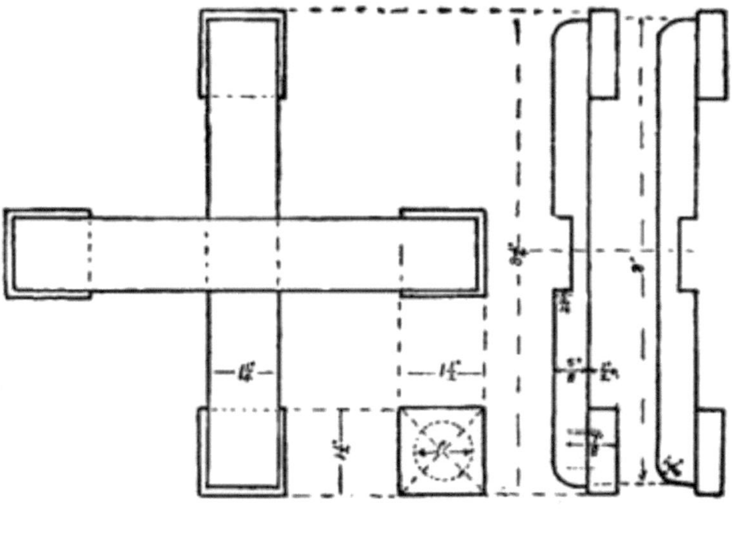

A pot-rest like the above is worth making because it lasts

The sawing and chiseling should be done carefully. It is necessary to saw on the inside of the lines or a loose joint will be had. Doubling the passage of the saw through the wood will often make the difference of ⅛ inch.

After these are made to fit, the upper ends may be rounded down by chisel and compasses, or bevelled, using the plane.

Use ⅞ inch brads or finishing nails, four in each pad or foot to fasten pads to the arms. The pads should project ⅛ inch from ends and sides. To finish the work nicely so the rest will both look well and stand exposure, apply a suitable stain. Allow it to stand at least thirty minutes. Then rub down with a cloth to an even stain. It is better to allow the stain to stand a day or so. This gives time for the stain to set before applying the wax. Otherwise, some of the stain will be loosened and removed when waxing and a lighter shade of stain will result.

PLANT JARDINIÈRE

The measurements are easy since the scale is one-half inch. That is, as you measure the line in the working plan you allow one whole inch for every one-half inch you measure on that. So, if a line measures three and one-half inches, make the line for your box seven inches. This is the real height of the box. Notice some lines have their real measures given at the side.

Directions for making are as follows:

Fold a piece of paper large enough for one of the sides and sketch one-half the outline on one of the folds. Cut to line and then draw the other half. This will give perfect balance. Cut two pieces of wood from this pattern by placing it on the wood and tracing.

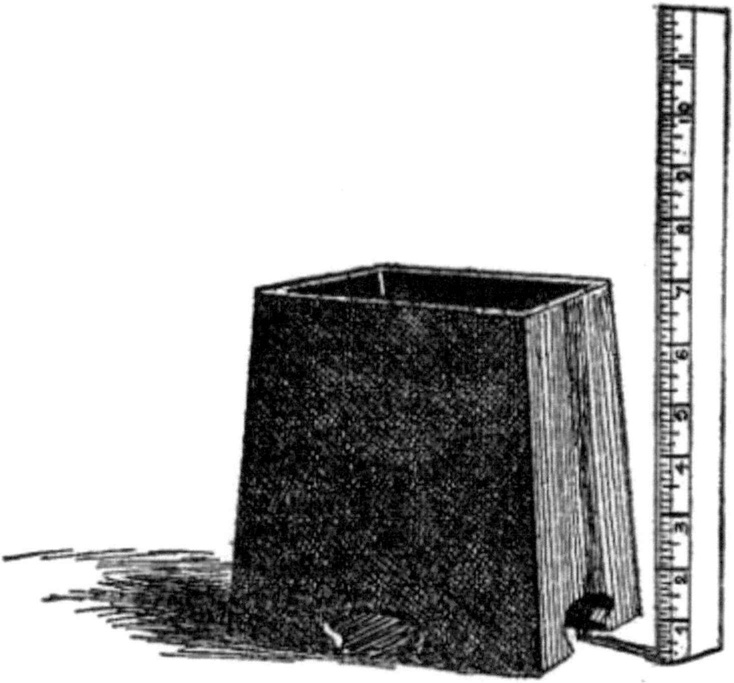

From the ruler you can read off the exact size of the jardinière

Draw a line parallel to each side ⅜ inch in on the pattern for a new pattern for the other two sides. These sides will need to be ¾ inch narrower, ⅜ inch on each side, as they must fit between the other two sides. If wood of different thickness is used it will be double the thickness.

Use a coping saw to cut out the base. The tapering sides may be cut to lines by saw, plane or chisel. The curve at the base may be bored by ½-inch auger, and in this way a better curve may be had.

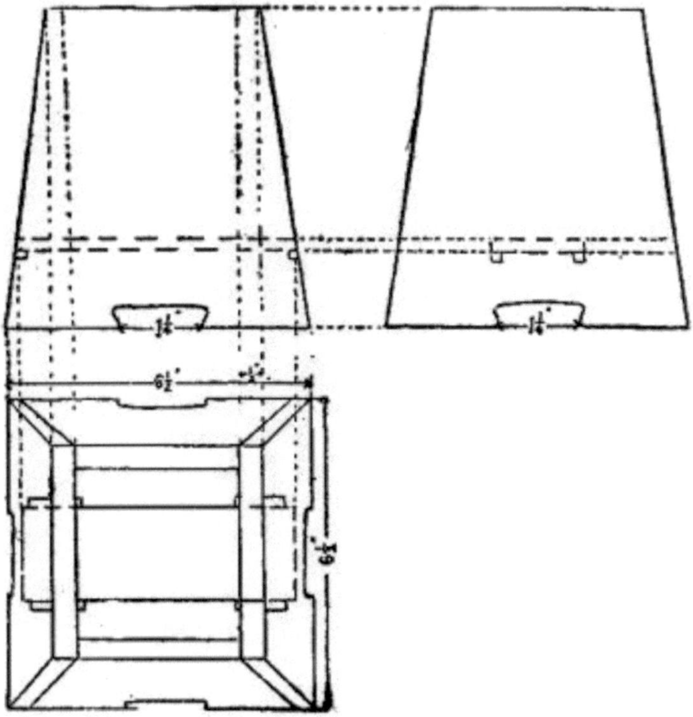

Helena used a scale of ½ inch in the construction of this jardinière So the plan is very easy to work from

Use 1¼-inch brads or finishing nails. A little glue added will make a firmer box. A much larger box after the same pattern will make a beautiful holder for a larger plant or shrub, using, of course, thicker wood.

Two small cleats should be nailed and glued from the inside to support a bottom. The bottom will give better service if it does not entirely fill the space. Let it be the proper length but allow a space of an inch on both sides for dirt and leaves to fall through and out.

Chestnut was the wood Helena used. It was stained and later waxed and polished. A beautiful permanent brown stain may be had on chestnut or white oak by applying strong ammonia to it with a brush and later sandpapering down and waxing. White wood is another good wood to use, but a stain will have to be applied to white wood, as ammonia will not act on it. A strong solution of permanganate of potash put on with a brush will darken any wood; it has no fumes.

VII

IMPROVING THE SCHOOL GROUNDS

During the first days of early spring The Chief and his boy assistants looked over the school grounds to see what should be done for its improvement.

The school was situated on a triangular piece of land right in the fork of two roads. The land was elevated; so much so that the building stood on a real slope; it was practically a road bank. This slope was washed by spring rains leaving large rocks exposed to view. The country road was especially poor at this section. There were deep gullies in it; the gutters were full of leaves and rock. About the school building was a comparatively level spot covered with rock. No trees grew here; a little grass struggled up each year, soon to lose heart and die.

"It seems to me," said Albert thrusting his hands deep down into his pockets, "that we have our life work here."

"Not at all," announced The Chief, "this is just the sort of thing which confronts most country schools."

Sitting on a rock The Chief gathered his clan in solemn conclave. At the close of the conference Jay marched into the schoolhouse and wrote the following headings on the board:

I. Constructing a wall to form an embankment.
II. Cleaning the grounds and making a lawn.
III. Planting of trees.

IV. Preparation and planting of the flower garden.
V. Cleaning and mending the road.

These headings represented the general lines of work the conclave had decided were the right ones, the most pressing ones to begin on.

First all the stones were picked up. The smaller boys and girls made little heaps of the small stones, while the larger rocks, requiring strength to move, were left to the older boys and girls. To some rocks the boys were obliged to take the pickaxe and crowbar. These were rolled, dragged and carted to the gutter at the bottom of the bank.

A sand bank of this description where the wash is great always needs an embankment of some sort to hold the soil in place. So the boys built a stone wall. They made this wall of the stones picked from the grounds. First the height was decided on. This was to be two feet. They drove stakes, one at the beginning, and so on for every five feet of extent. After leveling, two inches was measured from top of each stake down and a cord was strung along from stake to stake. Previously, to be sure that the stakes were at the same level, one of the boys, squatting down on the ground so that his eye was on a level with the stake nearest him, looked or "sighted" along the stakes. Where one stake seemed to rise up above the others it was hammered down a little to fall into line. Thus a straight line or top level for the wall was obtained. The wall itself was not difficult to build. It meant only the selection of stones and firming them into place.

Close to the wall there was a strip of level land; then the slope arose from this quite gently. After the stones were picked off the boys raked the ground all over fine, free from lumps and small stones.

One evening in the village store George's father offered to plough and harrow the entire grounds if Jack's father would give the grass seed. The bargain was sealed. But after all, this sandy soil was no sort of soil to plant grass seed in. The father of one of the girls gave to the school a few loads of good soil. This was spread over the slope to a depth of about a foot. Again they raked it all over smooth, filling in and making as pleasing a grade as possible.

The Chief told them it would have been far better if they could have had two feet of good soil. Grass needs all of that. Another way to have improved the soil conditions would have been to plant corn or potatoes on this ground for one year. With such a crop the boys and girls would have been constantly working it, stirring it up. This improves soil.

After the soil was spread the next thing was to make it firm. This was done in three ways. One day the teacher decided that for gymnastic work they might all turn out and tramp the soil. Up the bank they stamped, then down by the old drive to the road again, and up the bank.

Another way was by using tamping sticks. The boys made these sticks from old broom handles, to the ends of which they fitted solid pieces of board about ten inches square. Some were merely nailed upon the ends of the broom handle; but this method was insecure. The others were made with holes in the centre of the boards of the same diameter as the handles. These sticks were used to tamp the soil or spank it down. But on the day when an old farmer, stopping to watch the work, offered his roller, there was great rejoicing. Between classes, during recesses and at any odd time the slope was rolled. One boy in the very beginning pushed the roller but not after that, for when it was explained to him he understood why he should pull the roller. First, because pulled there are no foot prints left; and secondly, one slips and makes bad places on the lawn when pushing.

Next came the seed sowing. The allowance of seed was one quart to each 300 square feet. Jack's father chuckled when his son refused absolutely the variety he offered him. "No, sir, I do not wish Kentucky Blue Grass. It takes three years to get good results from it. The results are all right."

"Thanks," murmured the highly entertained father.

"We can't wait three years, we must have speedy results. I wish a recleaned mixture, and no chaff in it."

"Very well, young man, I wish to know two things: First, where did you get your knowledge? And second, where does my pay come in?"

"The Chief told me what book to read to understand about lawns. As for the pay, you made your bargain with George's father. Anyway I should think it would be pay enough to see a fine lawn in a public place made from your grass seed."

"Right you are, young man. Go on, read and read. But remember to work as well."

They chose a rather cloudy day for the planting, and a day when the wind did not blow. Grass seed is so fine it will blow all about if the wind is stirring. Grass seed is sown broadcast, that is, scattered by the hand. It is not sown in drills.

It was a pleasure to watch the sowing, for it was done right. First, the sowing hand was held low, the person stooping down. Some seed was taken with the fingers. Then the sowing arm was swung freely in a semi-circle. After going over the ground once, a second sowing was made at right angles to the first. A second relay of boys and girls came out and raked the sown ground all over. A third relay then rolled the ground. Do you see that there was little opportunity then for the seed being blown off the surface of the ground?

The children were delighted when a gentle rain, followed by several warm days came right after the sowing. A soaking rain or a series of cold damp days might have spoiled the work. The only way to have a good lawn from a poor piece of land is to do a thorough piece of work. Patching up means constant patching.

The paths and driveway to the school were just rock masses. The first thing was to clear out all the rock. Then loads of ashes were brought from the houses of the different children. All the parents were glad to get rid of the ash-dumps in the backyards. All kinds of carts were brought into use. For a week no boy dared appear without a load of ashes.

All these ashes were dumped into the drive and paths. Then the whole ash layer was rolled and rolled. It finally made a good solid kind of walk.

It was the business of the tree-planting committee to have two saplings ready by Arbor Day and to know themselves just how to plant. In the start of this work, committees had been formed. Now these committees were supposed to know exactly how to do the work and to procure the necessary material for it. It was not the duty of the committee to do all the work; by no means, or the others would not have known how to work.

Two trees were to be planted, one little maple near the building; another, a buttonball tree, down on the lower grade. A maple was chosen because it was easy to get from the woods and also because the maple is such a good all-round tree. Then later, because of a cold wind exposure on one side of the schoolhouse it was decided to plant a screen of little poplar trees. This was to shut off an unsightly view which could not be remedied in any other way.

One of the girls on the tree committee suggested a poplar in place of the maple. She was voted down. Now if quick results had been wished, of course the poplar would have been the tree to have chosen. That was why the poplars were chosen for screening purposes. But for permanence the maple, the oak, the buttonball are all better. The poplar shoots up quickly, to be sure, but again it sheds its leaves early in the season. Its life is not as long as the oak's. There are more reasons, too. But if you must have quick results, here is a trick. Plant first a poplar then a maple or some other tree and so on. Later the poplars may be cut out and you have left the fine sturdy, long-lived trees. At the same time the poplars have tided over that in-between period. We sometimes weary of waiting for an oak to grow sizable.

The tree planting was left until May because of the state Arbor Day. The maple and buttonball or plane-tree were dug up by the boys in the woods the morning of Arbor Day. The trees were chosen from a rather open part of the wood. It is better to choose trees from the open places than from the denser woods. Trees thus selected are far more likely to grow on being transplanted into a place similar to that from which they came. The boys chose trees about five feet tall. The smaller the tree the better. The following directions were the ones agreed upon:

(1) Dig a hole large enough and deep enough to accommodate the roots without cramping. Allow so that the tree shall sit one inch lower than it did before.

(2) Place the topsoil on one side of the hole; on the other the poorer subsoil. If the topsoil is very poor, get some good, rich, black soil.

(3) Place good soil in the bottom of the hole.

(4) Put the tree on this layer, spreading the roots out carefully.

(5) Shovel rich soil over the roots. See that it goes in between the roots. Don't be afraid to use your fingers for this work.

(6) The poorer soil goes in on top.

(7) Tramp the soil down with your feet, making firm about the tree trunk.

(8) If the planting comes late in the warm weather make the soil into a soft mud with plenty of water, in this form washing it in between and about the roots, all roots and rootlets come in direct contact with the mud.

(9) Last of all cut the tree back, shortening the larger branches about one-quarter their length.

After planting the boys kept the trees soaked with water, thus making it possible for the young saplings to have plenty of water. As the spring went on the little maple prospered but the plane-tree started to put out a few sickly looking leaves and finally died in midsummer. Just what was the trouble? Supposedly these two trees were planted according to the same directions. It finally came out that the boys who planted the plane-tree had not cut off the bruised rootlets. These rootlets being in a bad condition rotted and affected the entire root. Another mistake was the failure of the boys to put the good soil about the roots, and they had made the hole a little too small for the entire root area. Well, it simply went to show that such a piece of work must be done right and carefully, if success is to be certain. These were the reasons why our boys lost one of their Arbor Day trees. The Chief told the children that it might have been done over then, but that spring was the better time, because the transplanted tree has the good long feeding season ahead of it, and therefore has an opportunity to get over the shock and to get accustomed to its new surroundings before winter is on. Trees planted in the fall should not be cut back. Leave this until the next spring.

The children wished later that they had used something else for a screen. The poplar trees grew fast but of course did not fill out as evergreens and shrubs do. So, after all, the hedge of shrubs would have acted as a better screen. Had they chosen evergreens these would have made a better wind-break in the winter season for the exposure was north, cold, and windy. Such work, though, is worth while, because we learn so many better ways of doing things.

The flower garden was almost entirely the girls' work. In the first place the school had no money. Seeds do cost something. But the amount of seed which can be purchased for one dollar is amazing. Peter's grandfather, hearing of the school's needs, gave a dollar. This was money enough to buy seeds of ageratum, zinnia, dwarf nasturtium, California poppy and verbena besides some others. Most schools have interested friends.

All along the sides and front of the schoolhouse close to the building the nasturtiums were planted. The ground was hard packed. The plough had left the soil untouched near the building. So the boys spaded this up. All the stone was picked out. Good soil was brought from the woods, fertilizer from the barn and it was all worked thoroughly in.

Stakes had to be made. An easy stake to make is one from a lath. Mark off 18-inch lengths or such lengths as are required. Make one end pointed for about six inches; sandpaper. You have a good stake, that is, a good temporary one. These were driven in to the outer edge of these nasturtium strips at distances of four feet and strung with three cords four inches apart. The cords should be carried about the stakes in a groove made for this purpose. Thus the cord will be held and not slip up or down. Thus strung off, border beds will not be stepped on or run over by cats and dogs.

The nasturtiums were planted four inches apart, in drills one foot apart. Just two rows were planted. The first row was six inches from the front edge, then a foot space left, then another drill. Finally one foot was left between that and the foundation of the building.

The girls of the fourth grade made the drills with the hoe handle. The children of the first and second grades cut out pieces of paper in inch lengths. Four of these placed along in a row gave the right distance for planting the seeds. The nasturtium seeds were soaked over night. And since the soil was warm and mellow, it helped.

Along the walks ageratum was planted in the following manner to serve as a border. A drill was made as if for lettuce planting. The seeds were sown in the same way as for that vegetable. When the plants were an inch high they were thinned to six inches apart.

The zinnia was planted according to Helen's way of planting and told by her under the girls' planting in a later chapter.

The verbenas, as the other flowers, were planted in early May. They were planted one-fourth inch deep and six inches apart in drills one foot apart.

The poppy bed was made fine, very fine, by much raking. Then the seed was sown as the grass seed was, that is, by the method we term broadcast sowing. These plants were thinned later so as to stand about eight inches apart. But the plants thinned out were not used again, for these poppies will not stand transplanting. This bed was simply one gorgeous red in August.

In the early spring days the gutters were cleaned out thoroughly. The road patching was quite a different matter. These country roads, like those of many places, were just dirt roads. Now earth is poor material for road construction. But if drainage is properly looked out for, and the earth road is smooth from rolling, earth roads make, after all, fine roads for summer travel.

It was suggested that rock be filled in, and the earth over this. But when the boys considered how deep cuts would be formed in such a mend by wagon wheels, this was given up. Then it was decided to fill in with layers of rock mass. Myron brought a load of slate for this purpose. But slate, while it makes a smooth road, does not stand wet weather well. So Myron had to return his slate to the road-side bed from which he had taken it. Then The Chief told the children briefly about road materials; how soft limestone makes too weak roads for loads, how easily they wash and wear; how granite, because of its being made up of several materials, is poor, too; how flint and quartz, while hard, are brittle, and are not sufficiently tough; and that sandstone was impossible. Then he told them that good gravel, tough limestone and trap-rock were good road materials. Roads need materials having hardness, toughness and cementing qualities.

By taking a trip to a gravel bed, some three miles out of town, the boys were able to get gravel for their patchwork. They did not merely fill in the breaks but dug out the road bed straight across wherever a break occurred until they came to good road. Coarse gravel was put at the bottom up to six inches of the top surface. This was packed down and rolled. At the same time it was watered until mud rose or flushed over the top surface. Finally pebbles from about a half-inch size to coarse sand were laid on and rolled thoroughly.

This is the way these lads fixed one piece of poor roadway.

It happened that one of the farmers near by tethered his cow on the school grounds during the summer. One of the girls gave a workable solution for this problem. This was it: the boys should come back in relays all summer long and keep the grass so short that no cow could get a nibble from their new lawn. This was done and it worked.

When the subject of the care of the flower garden arose it was easily settled. The girls gladly divided themselves off into committees. Each committee's business was that of weeding, picking and distributing the flowers. The prophecy that there would be blossoms enough to supply the homes, the churches and the sick proved true. To be sure the garden did not look so well in the fall as in early summer, but it took only a short time to fix up the grounds when school re-opened.

Plans were made for another spring during the first weeks of school. The lawn would need a little more work done on it, an oak should be planted, a group of shrubs put in. But the foundation work had been done.

And one day when the news was brought that the town was going to put the first strip of real macadam road by the schoolhouse, a deafening shout went up.

VIII

MYRON'S STRAWBERRY BED

One fine day in early April Myron spaded up his strawberry bed. The bed was made in a sunny spot, on moist but not soggy soil, land excellent for strawberry culture because the year before it was part of a potato field. Following The Chief's advice he had spread over the bed only a very light covering of well-rotted manure. Myron first measured off his garden bed driving stakes in at the four corners. Then he strung off the bed with stout garden cord. "Now," he said to himself, "I know exactly what I have to do." Then going to one corner of the space with his back toward all the rest of the bed he began his work.

Photographs by Edward Mahoney

The Way the Chief Taught His Boys to Handle Tools

He had a fine spading fork which he had bought a few days before. Grasping the top of the handle with his right hand, with the left midway down the handle, he pressed the prongs of the fork with his left foot vertically into the ground. Then lowering the top of the handle toward the ground and backward, he slipped his left hand down the handle to about a foot from the prongs, and drew up the spading fork with earth on it. This earth he threw a little forward and with the prongs broke up the lumps. He continued this until all the work was done.

Then he looked at his spading fork, his brand new fork, and found the prongs quite bent, "The Chief told us to buy decent tools, but I thought I'd save a little money. Well, I'll break up some of these lumps a bit with my hoe and see how that will stand a little work." The land Myron's father had given him was very good indeed, rich and light, so that work of lump breaking was really very slight, yet it made the new hoe-blade rattle in its socket.

After this work had been thoroughly done the boy took his rake and started making fine the soil for the bed. Myron had learned well how to handle his tools. These lessons of handling tools The Chief had taught the boys for he felt that a tool should be a skilful instrument in the hand. "A gardener should wield his hoe as well as a surgeon does his scalpel," The Chief had often said. So the boys were proud of really knowing how to work.

After looking proudly at the fine, smooth bed the boy shouldered his tools and marched off to the village.

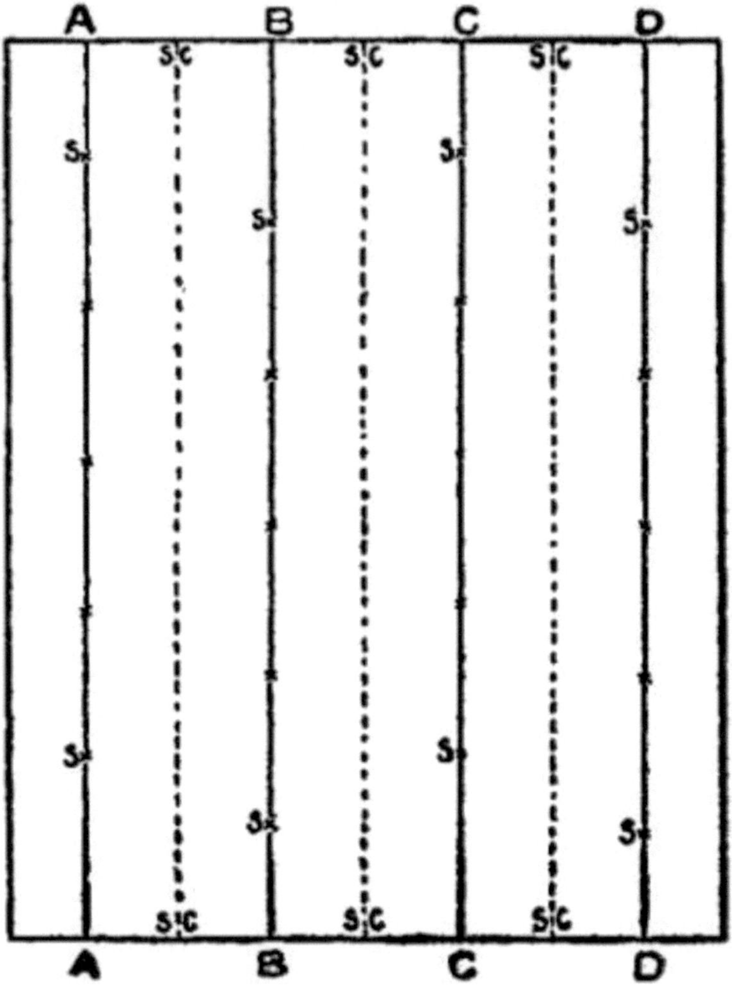

The crosses show where Myron set the strawberry plants. The dotted lines signify the plantings of succession crops

Do not think that you can save money by purchasing poor tools. It is quite impossible, because always one has either to buy new and better ones, or mend and remend the poor ones. The lad found out that a good trowel costs at least 50 cents although a smaller one called a transplanting trowel may be had for 15 cents; cast steel rake, 50 cents (10 teeth), 75 cents (14 teeth); hoe, 50 cents; Dutch hoe, four inches, 40 cents; spading fork, $1.25, and weeder 10 cents.

That afternoon armed with cord, stakes, a tape, and the plan of the bed, Myron started to mark it off for the plants. After tacking his plan up on the fence post he began the measuring. The piece of ground was 5¾ feet wide by 6 feet long. Beginning at one edge of the garden he measured in six inches along the width. The same thing was done from the opposite edge. Stakes were driven in at these two points and a cord stretched between. The same thing was done from the other two ends. So Myron had two cords extending down the length of his garden each six inches from the edge of the patch. These cords are lettered A A and D D in his plan. B B is 15 inches from A A; C C is 15 inches from D D.

The next thing was to get the position of each plant in the bed. This is the way it was done: beginning with A A, measure from the upper stake nine inches down the line and place a small stake. This is the place to set the first plant. From this, measure and place stakes at one-foot distances. There will be five plants down the line. Down B B, measure fifteen inches and place a stake. This gives the position for the first plant, then, as before, place stakes at one-foot intervals. C C is marked off similar to A A; and D D to B B. In all Myron then had places for twenty plants.

As the work was finished Myron looked up to see Jack's face peeping over the fence. "How do you like my strawberry bed?"

"It's all right," Jack replied, "especially the strawberry plants. They look very promising."

"Quit your fooling, and come in and see this bed face to."

As Jack went over the fence he stopped to look at the plan. "I say, Myron, this shows a plan's of some use to a man. What do you mean by succession crops?"

"That stands for the sort of seed you keep sowing at intervals and so getting several crops a season. I shall put in radish and lettuce. I am to supply our own table all summer. Father is not going to sow either of these. He is depending on me."

The trip to Longmeadow Farm for strawberry plants was one of pleasure and profit to Myron. The boys used to say that while old Mr. Mills had a crust inches deep, underneath this he was as fine as the strawberries he raised.

"Strawberry plants are worth," said the old gentleman, "about two cents apiece. I will give you your plants if you will do two things. First, during this season, you are to pinch all the blossoms as they appear, off the plants. Secondly, I wish to experiment with a new variety of berry to see if it is good for this locality. I wish you to take five of these plants and try the experiment with me. Do you agree?"

"Certainly. But can't I leave just one blossom on each plant to see what the fruit is like?" And also leave one entire row blossoming as it will?"

"Yes, that will be all right. The reason for pinching off the blossoms the first year is to save the strength of the young plant. Otherwise it all goes to fruit forming. It pays to do this, because the second year you will have a good yield. Remember that strawberries which flourish in certain localities may fail utterly in others. That is why you and I are experimenting with this new berry. I am going to give you five plants of Marshall, five of Nick Ohmer, and five of Brandywine. Remember, shorten back the roots three inches before you plant. I shall be around to see your strawberry bed. Remember to cultivate after every rain, and in between times, too."

"Thank you, and good-bye," said the boy.

Myron set his plants after the following fashion: he dug trenches along the cord lines previously marked out. Then the roots were shortened. To plant, hold the plant against one side of the trench just as Myron did, as illustrated in one of the pictures. Then push the earth in from the other side and press firmly in place. The plants should sit so that their crowns are even with the top of the ground. When Mr. Mills came to see that bed he found two or three plants badly placed. Care must be taken in the placing. The days after planting were very hot so Myron covered the plants with straw to protect them from the heat.

As the season advanced the little plants sent out runners. These were immediately cut off. If they had not been, they would have become entangled and thus formed what is called a matted row. Some people cultivate strawberries this way. But Myron's way, the hill culture, while it means constant attention, is perhaps a better method.

One day, old Mr. Mills took Myron on a little trip with him to a farm where a man was cultivating berries by the matted row method and doing it in a very slovenly way.

"It taught me a lesson," the boy told his mother that evening, "that lazy methods are pretty bad."

Once or twice that season he sprinkled wood ashes on the ground of the bed. Just a little should be sprinkled on, as one sprinkles salt on a potato. Soil gives food to a plant. This food is nitrogen in various forms, potash and phosphorus. Sometimes we help the soil supply one or more of these chemicals. The wood ash adds a little extra potash which is very good for the strawberry.

It turned out after a second year that the new variety gave very small and flavourless berries. So the old gentleman and Myron wasted no more space on that variety.

The second year Myron obtained excellent results. From some of his plants he got one quart of berries each, during the season. That was good, but no better than a strawberry plant should do under good cultivation.

As far as his lettuce and radish went there was nothing new or startling in his experience. He tried this little trick of lettuce sowing with some success: Instead of sprinkling the seed in the drill, he placed each seed separately and four inches apart. By this method one need not transplant to get good heads. He tried the Black Tennis Ball seed. This forms a good head.

Did you ever try the Icicle radish? Myron recommends it. It is long and white and so gets its name. Along with the radish he planted parsley. This is a good way to do as these vegetables do not interfere one with the other.

"Grow any more lettuce and radish?" exclaimed Myron's father one evening in the village store, "not while I have a boy who can do it as Myron can. He beats me all right. And I am glad."

IX

JACK'S ALL-ROUND GARDEN

Just as soon as the ground was workable Jack set his coldframe. He chose a southern exposure, back of the barn, so that the frame should sit up against the stone foundation of the building. First he dug down about a foot deep. As he dug, he knocked up the lumps and picked out the stone. Then he went to the barn and got a barrow load of horse manure, not fresh, but old, rotted manure. This he very carefully mixed in with the soil already made fine.

"Now I shall put the frame on. Come, Elizabeth, and give me a lift with this." After some tugging the frame was set.

"I thought frames were usually sunk in the ground," commented Elizabeth.

"I shall do that this fall and make a real hotbed out of it. You see this spring I just want to give my seeds a little extra start. That's why I made the soil so rich and so deep. Now I am going to bank the frame about with manure. Then I shall put dirt over that. You see I get some extra heat that way. Just see the fine slope of the glass. I guess Old Sun will get caught all right."

Jack busily banked the frame, spanking the fertilizer down hard with the back of his spade. He sloped it up some four inches along the sides and front.

"Now I am going to make drills for my seed. In the first partition I shall plant lettuce and tomato; then pepper and onion go in, and the third is for flower seed." Jack bent over the frame, and began to scratch lengthwise of the beds with the edge of his trowel. Red-faced from bending over, and hot from his former exertion, his trouser knees covered with earth and manure, he stood off and looked at his work.

"I'm precious glad Elizabeth has gone, for if those aren't the worst, crookedest old rows I ever saw."

And so they were. They were all distances apart, of different depths and entirely untidy-looking.

Jack picked up his rake and again raked the little beds over, so that no trace of his poor work was left. Then he found a board which stretched across the frame widthwise, so that he could kneel upon this and work to advantage in the bed. He next whittled out two little pointed sticks to act as stakes, and tying to these a piece of cord just the right length for the drills, he was ready for work. With one stake stuck in the bed at the upper end, the other at the lower, the cord between gave Jack a good string line for the drill. Then, with the end of a small round stick held close against the taut line, the drill was made. So he continued making drills at distances of four inches apart.

Pouring out some lettuce seed in his hand, Jack began to sprinkle it rather unevenly in the first little drill. Elizabeth, having returned, stood by watching and shaking her head. "I didn't know you were here. You make me nervous," began Jack.

"I feel more nervous than you possibly can, for you are wasting seed and sowing in a poor way. See, here you have a little pile of seed, and there you have none," and Elizabeth bent eagerly over the bed.

"Well, if you think you can do better, just try this next drill." Jack straightened up, and gave way to Elizabeth.

"Wait a minute," and Elizabeth ran into the house. Soon she came out with some small seed envelopes in her hand. From the bag of lettuce seed—for Jack had bought his seed by bulk—Elizabeth poured some into a small envelope. Then by shaking the envelope she carefully and sparingly sowed the lettuce in the drill.

"I say, that is good!" said Jack admiringly. "Now I'll do some myself."

"I should think you would wish only one more row; then have a row, or perhaps two, to transplant in. For I believe you'll have to prick out the plants before the garden is ready."

"You talk like the real thing, Elizabeth. What do you mean by pricking out?"

"Why, pricking is just lifting out the seedlings with a pointed stick from one row to another, or from a box or hotbed into the outside garden. What else are you going to plant, Jack?"

"I thought I'd put in—say two rows of tomatoes, one row of onions, and one of peppers. In the third partition I'd start asters. I just love asters. So I've made up my mind to make a kind of specialty of these."

"That's fine! May I help?"

"You certainly may, for you are a help."

Elizabeth chuckled away to herself, for Jack evidently was not questioning where she got her knowledge. "It seems to me," she rather timidly suggested, "that it would look more shipshape to label these rows, and put in little sticks where each row begins and ends."

"Well now, that is a fine suggestion." So Jack stuck in some little sticks he got from the woodshed. Elizabeth did not dare offer some nicely made little markers laid away in her desk for future use. She feared those would call forth questions.

Jack brought out a hammer and tacks. Then writing the names of the seeds on the little envelopes Elizabeth had brought out, he tacked one over each row onto the inside of the frame. They both stood off and admired the work.

Warm days Jack opened the frame, at first only a little, and later, wide open for all day. One night he forgot to close it, and a slight frost made a sorry looking set of seedlings next morning. He lost every single plant except a few little asters, which were protected by the inner partition of the frame. These seedlings he watered at intervals all day. This was at Elizabeth's suggestion. By this treatment these were saved. So Jack, sadder and wiser, started over again.

When the lettuce plants had four little leaves Jack, with Elizabeth's help, transplanted some into the drills left for them. When they were larger yet, they transplanted the lettuce to the real garden. This is the way they did it. In the first place the children chose a cloudy day for the work. A cloudy day is far better than a bright sunny one because bright sun is too strong for little lettuces which have been disturbed from their places and put into new ones.

To transplant, dig up a number of plants and plenty of earth with them. Use a trowel for this work, gently lifting plants and earth. A drill may be made; or, perhaps better yet, make holes with the dibber. Pour a little water into the hole. Then gently separate a plant taking as much soil with it as you can keep on its roots. Place the little plant in the hole or drill, and cover the roots with soil. With the fingers press the soil firmly about the plant. Water the earth, not the leaves of the plant. Next day, and for several days, cover the transplanted plants with strawberry baskets. These are far better than newspaper coverings, because light and air freely come through the crevices of the basket. The newspaper makes a covering too tight and close for the tender lettuces. Between plants the children left six inches.

Jack raised Boston lettuce. He not only had enough for his mother all summer long, but sold some, too. The way he happened to sell it was merely an accident. Not far from the village was a large summer hotel. One day the proprietor had driven around to the house to see Jack's father on business. As the men were talking Jack and Elizabeth came from the garden with two fine heads of lettuce.

"Have you any more lettuce than what you can use yourself?" asked the proprietor, after feeling of the heads of lettuce and admiring the good firm centres. "Yes," replied Jack, "I have now, and shall have all along, more than we can use. You see I keep making sowings every ten days in the coldframe, and transplanting."

"I'll take all the extra lettuce you have at five cents a head. That is what I pay all summer long for it. To-morrow bring me up what you can."

"Thank you, sir. Ten heads will walk up to-morrow."

"The first time I've ever heard of heads walking," laughed Jack's father, well pleased with his lad.

But we are away ahead of the story, for we have planted and sold lettuce before Jack has had a chance to really make his garden. The soil in the backyard was very poor, so Jack decided to cultivate only a strip twenty feet long and eight feet wide. He dug out all the soil to the depth of two feet. His father lent him the use of a horse and wagon, and gave him from the barns whatever fertilizer he needed. The digging was a long, tedious piece of work. It was hard, too; but the boy kept at it. Any piece of land can be used if a boy has a mind to work hard over it.

Some of the poorest of the soil was carted off, then into the top of the remaining soil he mixed the old manure. Then into the garden space six inches of manure was spread, and over this was filled in the old top soil and fertilizer, that mixture which he had previously prepared. About one foot of this was put in. Jack's father lent him the horse again and the services of a man. They drove to the Longmeadow Farm and got a load of top soil. Old Mr. Mills said he would give the soil if Jack could answer three garden questions correctly.

"All right," said the boy, "you'll probably knock me over, for I don't know much about gardening, but I'm trying hard."

"Question number one: suppose your backyard had been clay soil—what would you have done with it then?"

"I should have mixed in sand, using about one-quarter the amount of sand as I had of clay."

"Good! Question number two: suppose you had no sand—what then?"

"I'd have used ashes; old clinkers I guess would be best. Everyone has ashes."

"Question number three: what is the object of mixing sand or coal ashes or clinkers with clay."

"The reason is to break up the clay. Clay bakes hard, becomes sticky, and little air or light gets into it. Ash or sand breaks it up. I think that's about all I know about this."

"The soil is yours, young man, I shall be around to see your garden some day. Remember good gardening means working your muscles hard."

"Thank you, Mr. Mills. By the way my arms and legs ache, I guess I know about muscles."

"And remember too," continued Mr. Mills, "that certain vegetables are very closely related and will intermingle. For example, do not plant different kinds of corn close together. The pollen from one kind will fertilize another kind and so you get a crossing which results in a mongrel sort of corn. Melons and cucumbers will do the same thing. And so care must be taken in order that this sort of intermingling does not take place. You see, Jack, that there are many things a real good gardener has to consider. Gardening is not only a matter of soil preparation but it is also a matter of understanding plants and their relations one to the other."

So the good soil was put on and Jack was ready for business. Straight across the back was planted a row of sunflowers. Sunflower seeds belong under the head of large seeds, and should be planted one inch deep and one foot apart. Two seeds were placed in together. This is a safe plan, because if one fails to come up, the other doubtless will come up. If both appear, when the plants get about three inches high, pull out the weaker one.

Then the boy planted a second row two feet from the first one. The first row was planted close up to the fence. Jack found out that this was a mistake. Always leave all about the garden a space of a foot or so, in order that one may walk about freely and get at the rear row of plants without trouble. Again, do not plant too close to a fence, unless the planting be some vine or climbing plant, which you desire to have cover the fence.

Next the aster plants were transplanted. This was done after the same manner as the lettuce. They were placed about one foot apart each way. These were put across the entire spot just as the sunflowers had been. Thirty-two little aster plants were set out and still Jack had a number left over. It is amazing the amount of aster plants one can raise from a little packet of seeds. "I'm going to sell the rest of these aster plants," he declared. And he did. The boy tramped about until he found a lady desiring the plants, to whom he sold 50 little plants for $1 and set them out for 50 cents.

The rest of the garden space was used for the onions, peppers, lettuce, tomatoes and radish.

The onions transplanted from the coldframe gave fine early onions with a mild flavour.

When Jack was making furrows for the sunflower seed Jay came along and leaned over the fence. "Jack," he drawled, "you look like a kangaroo all humped over making that furrow. Why don't you use your hoe right?"

"I thought I was using it right. Come in here and show me how, will you?"

So Jay jumped the fence and picked up the hoe. "Stand this way! Straddle the furrow with your back in the direction you are going to hoe; or else stand on the left side of the furrow facing it. Grasp the handle of the hoe in the right hand near the upper end. The back of your hand should be up. Now the left hand should be a foot or more below the other hand. And see the back of my hand. It is toward the left and my thumb points down the handle, just so with the rake handle."

All summer long the boy worked or cultivated his piece of land. He kept hoeing and weeding constantly.

One of the August pieces of work was to fix the hotbed for winter. Now the frame was taken up and the pit dug deeper—about two feet this time. Previous to this a great pile of manure had been heaped up near by. Jack had sprinkled it with hot water to start fermentation. Steam rising from the heap was proof of this, and it may be used at this time.

Then the manure was put into the pit. An eighteen-inch bed of it was made and firmly tramped down. At first the temperature of this was over one hundred degrees. When it dropped to ninety-five degrees soil was put on. The temperature was taken by means of a thermometer buried in the manure. The frame was placed after two inches of soil had been put in; then four more inches went on. The surface of the soil was made to slope at the same angle as the glass. All about the frame was banked, again, manure covered with earth and leaf matter.

Jack transplanted violet plants into one compartment. These were good violets and were placed four inches apart. In the second bed he sowed foxglove, pansy and stock. The third was left for radish and lettuce, a bit later.

Elizabeth helped him sew together several thicknesses of straw matting as covering for the winter nights. They had decided that newspapers next the glass, then the mats, and finally a rubber blanket, would be protection sufficient.

But Jack's hotbed work is quite another story. However, I can tell you that the next winter he added two other frames to this one.

X

ALBERT AND JAY'S DRAINAGE PROBLEM.

The problem of draining which Albert and Jay had to consider, was perhaps the biggest piece of work that was done all that spring. In the first place, it should have been done in the fall. That is the time to do such work, for if put off until spring it delays greatly the spring planting.

It was a wet spring, too. The boys, rather impatient of waiting, started digging one day, but it ended in disaster. The ground was soft and wet and hence very heavy to handle. This piece of land was one hundred feet wide or deep. It had a frontage of one hundred and fifty feet. A slope rose up in front of it, which accounted for the water being drained onto this land. The water naturally would have run off the land into a brook at the back. But in about the centre was a hollow, and beyond that the ground rose a little, and then dropped toward the brook. The depression made a kind of drain hole and the water settled there all the spring through.

This strip of land of the boys was not by any means the entire piece of land, which was much larger, but the boys' father had given them this largely to try their mettle. He felt so certain they could not do it that he said they might have all they needed from a pile of drain pipe he intended to use himself on a piece of wet land the next fall. "I shall have all my drain pipe left to me," he said to the boys' mother one night. She smiled, for the boys had talked matters over a bit with her.

Myron's strawberry bed was all made, Jack's garden-filling work done, George's ploughing and planting finished, before the boys could lay the drain.

"It's no use," said Albert, "I'm ready to give up."

"Now Savage, there's to be no quitting. I'd be ashamed of you, at least we can surprise father."

"All right, Jay, I'm with you."

Finally the day came when The Chief and the boys started work. A drain pipe should be laid ordinarily anywhere from twenty inches to three feet deep. One may dig or plough to make the trench. It is wise to dig as narrow a trench as possible and so lift as little soil as possible. Then, too, the bed of the drain should slope gradually from the upper or highest point to the lowest. The drop in level should be about four inches per hundred feet. So the boys had to consider just this. This is the way they "sighted" to get the drop in level. They drove a stake into the ground at some twenty feet from the place where the drain was to begin. Previously a cord had been stretched from one end of the centre of the field to the other end. Since the centre of the field seemed to be the place for the deposit of water the drain was to go directly through the centre.

If you ever have a piece of draining to do the problem may not be so simple as this. You may find several natural drainage areas. Then you must lay drains through these. Or instead of separate drains make side ones which empty into a main drain.

Going back again to the "sighting" for the drain bed level—the boys have driven a stake into the ground. It stands five feet above the ground level. If a tree had been in line with the drain line this might have been used and saved driving the stakes. Across the stake, at right angles to it, a board with a perfectly straight edge was nailed. This board was about four feet long, one end pointed at the drain line. At the other end Jay placed his eye looking across this to where Albert had driven stakes.

One stake had been driven into the ground at the beginning where the drain was to be dug; another at the extreme end or outlet of the drain. Albert stood at the first stake and ran a little piece of paper slowly up and down the stake until Jay raised his hand. This meant that the paper was on the same line with the sighting board. Then Albert ran to the other stake and did the same. The difference in these two points gives the difference in level of the ground. Albert measured from the ground to his mark on the first stake, and, doing the same in the case of the other stake, found the difference to be eight inches. This was too great a drop. Then the boys drove two stakes in between these others and did the same work of level finding. From stake 1 to 2, or for the first twenty-five feet there was no difference in level. For the first fifty feet there was four inches drop; for the next twenty-five feet, five inches rise; and the last twenty-five feet, six inches drop. They marked all this on the stakes in order to make sure they got the level right. The bed must, you see, drop one inch for every twenty-five feet. For the first fifty feet of the line the drop was just twice too much; then came the abrupt rise and drop.

Albert ploughed a furrow straight along the line and ploughed back again. Then he reploughed. The boys then began to dig, making a ditch three feet deep right through the land. In order to get the right level they used a home-made device and plumb-line which can be made as follows: Nail the ends of two six-inch boards ten feet long, so as to make a right angle; then across the open end of the triangle, nail another six-inch board having the lower edge about a foot from the ends of the boards. Cut off the ends of the boards on a level, so that they will rest evenly on the ground. Next drive a nail into the apex of the triangle, and to it tie a line long enough so that when the triangle is stood on its legs, the plumb-bob, which you will tie on the other end of it, will almost reach the ground.

The centre must next be determined. To do it, set the triangle up on its legs on a level place and when the plumb-line comes to rest, mark the place. A lead-pencil mark will do, but as it is liable to become obliterated by the dirt, a saw mark is more permanent.

Now you know what the grade of the bottom of your ditch will be. Reproduce this on a level place by means of a board with a large enough block under one end to give the right pitch; put the triangle on this and when the plumb-line comes to a rest, mark the place on the cross piece. Reverse the ends of the triangle to get a similar mark on the other side of the centre or level mark. This makes a level by which a fairly accurate grade can be made.

The tile pipes were laid upon a bed of gravel. This prevents the clogging up of the loosely put together joints. To fit tiles place the small end of one into the large end of the next, and so on. Over the end of the last tile, which emptied into the brook, they wired a bit of rather fine-meshed chicken wire.

Then the trench was filled in. By test Albert had found the soil of this land acid. Lime was to be put on it. Now lime must be in a crumbling state for this purpose. So after they had bought the lime they dumped it in a heap on a corner of the plot. After it had become air slaked, or reduced to a powder by the action of air upon it, it was spread over the lot. This and considerable fertilizer was ploughed in. The boys then had an ideal sort of planting soil for almost anything. The drain actually worked.

Now some boy may ask, suppose a fellow has no tile and cannot afford to buy any. In such a case there are two alternatives or choices. A wooden trough may be made by nailing together boards six inches wide. Then make a gravel bed and tip this trough over on it peak up. The wooden drain, however, is likely to rot. The other way is to put a double row of stones right through the centre of the bed slope. These stones—perfectly flat ones—should be placed on end with a foot between the rows. In this space put small stones.

The chief thing to remember in the drainage problem is that one wants a gradual flow of water from inlet to outlet. Any boy can fix his wet, soggy garden spot even though he has no tiles. Stones can be found surely, and, if no gravel beds are near, all the little brothers and sisters can pick small stones.

The boys had decided on planting what they called a general vegetable garden—corn, pumpkins, beans (bush), melons, tomatoes, beets and carrots. This combination of vegetables was a happy choice as they all can well wait until rather late for planting. The boys used the planting table that Peter had worked out.

Photograph by H.L. Schultz

Jay's Tile Drain Converted a Swamp Into a Garden

Many times boys and girls are bothered by the word "hill". I have seen boys make nice little heaps of earth and then make a hole in the top of these like a crater in a volcano. Down into this crater they poke seeds. Now a hill merely means a place. This place is not to be heaped up above the level of the ground. Place five seeds to the hill. Do not, of course, make a little pile of these seeds but lay them on the ground with a little space between seeds, say an inch. When planting beans place the eye of each bean down against the ground. The bean stands up on edge.

The pumpkins were planted in between the hills of corn. This is just the place to plant pumpkin or squash because there is opportunity sufficient for the vines to run. Remember not to plant these two together.

Soon after the pumpkins began to come up the boys noticed one morning that the leaves had been eaten. Some were completely riddled, looking like lace work. Digging about the ground Albert found a black and white striped beetle. Its name is the Striped Beetle. The boys killed these in the course of three days. They bought five cents' worth of white hellebore, which is a powder, and sprinkled it on the ground in a circle about the stems of the young plants. They made the circles some six inches from the plant stalk. Doing this at night, the evening dews prevented the scattering about of the powder. They put this on for three nights. Afterwards sand was sprinkled lightly over the hills and at the end of the runners. This makes a discouraging sort of prospect for the beetle who is hunting for something good to eat, not sand to walk over. If instead of sand they had used lime it would have been better. For the lime is quite likely to form a sticky mass on the legs of the insect pest. The moisture from dew or rainwater helps this along, while sand is far more likely to drop off the victim's legs. The Chief felt sure that besides the beetles there were slugs in the garden. Slugs are very likely to bother. They appear early in the season, feed chiefly at night and after rains, and lay eggs throughout the summer and autumn. These eggs are laid in the ground and in rubbish heaps.

The treatment suggested above and started just as soon as there are signs of slugs, will work. The boys treated their melons in the same way and had no further trouble with beetles and slugs.

The tomato plants were started inside. They were transplanted into strawberry baskets. These are excellent to use, because in transplanting to the ground the little strawberry baskets may be knocked apart without disturbing the plant nearly so much as if it were planted in a compact box. Be sure to line the basket with paper before filling with earth. When the plants began to straggle about and bend over stakes were driven into the ground and the plants tied to these.

Jay used hoops and made a sort of cage for the separate plants. He drove four stakes into the ground at distances of eighteen inches from the stalk and in a circular form. Then slipping hoops over the plant he nailed these hoops to the stakes. Some plants had two hoops about them, some three; it all depends on the size and needs of the plant. Only keep this in mind; that the object in staking tomatoes is not only to keep the plant erect, and the fruit off of the ground, but to allow plenty of light and air to get at all parts of the plant.

The bean culture resulted in a little private contest between Albert and Jay. That winter The Chief had given the boys a talk on inoculation of soil. One day while they were working on their land Jay suggested that they separate the bean section of their garden, having a bean plot at one end and another of the same size at the extreme other end; that one of them should inoculate the soil of his plot and the other should not. These plots being so far removed would not be in danger of soil washing one from the other. Albert, who rather scorned inoculation of soil, willingly agreed to make the experiment, stipulating that he have the uninoculated plot.

By inoculation of the soil is meant introducing into the soil a germ. This germ makes it possible that the nitrogen already in the soil be given to the plant in such a form that it may be absorbed, and absorbed in greater quantities than it otherwise could be.

Jay sent to the nearest State Agricultural Experiment Station, asking for the soil. This was sent free of charge. It was a soil, fine in texture and brown in appearance. According to the directions sent with it Jay spread it evenly over the top of his bean patch. A piece of land for inoculation should be prepared all ready for planting; then the inoculated soil is merely put over this, as frosting on a cake.

After this the seeds are planted. They planted bush limas. Of course they had to plant the same kind of bean for the sake of the experiment.

Beans are not hard to cultivate. They should be kept free from weeds and the soil well stirred up. Albert, fearful of his beans becoming affected by spots or anthracnose, sprayed them from the start. This disease is likely to affect beans about July. So in order to get ahead of the inoculated crop the boy did what he later found there was no need of. To be sure beans are liable to this trouble, but it is not a surety. It is never likely to appear unless the weather be very moist. This summer happened to be a dry one.

The spray he used was the Bordeaux mixture. His father offered to supply him with the mixture if he would do mixing for both. So he used this receipt: Dissolve six pounds of copper sulphate in six gallons of water. It is an excellent plan to crush up this chemical in a mortar and put this powder into a bag. Hang the bag up so it just touches the surface of the water. Add twenty-five gallons of water to this. To four pounds of slaked lime add twenty-five gallons of water. Then add this solution to the other.

The boy's father had a spraying machine. So Albert used this. I have known boys to use a corn broom to spray with. Dip this in the spraying mixture and shake over the foliage. The only spraying rule Albert used was to keep the foliage covered with the mixture; this does not mean many applications.

At the close of the bean season Jay had the finer, larger, beans with a better flavour. His yield was one-third greater than Albert's.

"And think, too, how I worked," Albert moaned. "Hereafter I shall not make fun of inoculation."

There is not much more to tell of this garden. The poppies yielded well. These were supported as they grew by stakes, as tomatoes are. Carrots need rather mellow, upland soil. The boys found that their carrots did not do so well as the other vegetables. The soil was a bit heavy and moist for them. They found this out about beets: beets should not be transplanted. Transplanting puts them back. Albert transplanted a few and learned this fact.

XI

GEORGE'S CABBAGE TROUBLES

George had a long task in stone picking. The old slope seemed to be full of stone. George would pick continuously from school to supper time, and next morning declare that new stones had grown in the night.

The ditching was very little work. It meant digging a ditch about two feet deep and then making at either end of this gutter a side ditch at a very severe angle to the main ditch. These side ditches were directed along the sides of the hill for about six feet, and the water thus directed would conduct itself off. Of course the angle was such that the ditch led away from the garden spot.

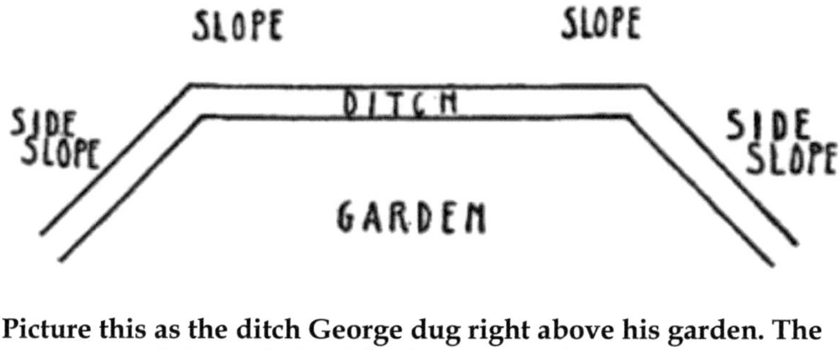

Picture this as the ditch George dug right above his garden. The water passed through the side slopes away from the garden.

As the stones were picked off he piled them into the gutter, where this stony bottom also helped the drainage problem.

George was a master hand at ploughing, for he had always done his share of it, so ploughing meant nothing to him. First, you will remember George had one foot of dressing to put on the land. This he ploughed in; and then reploughed. After this the slope was harrowed. You all know that the harrow simply makes fine the soil after the plough has done its work of throwing up the earth. The rake is a kind of harrow. Of course, when the garden plot is large, the rake is impossible, and then the harrow, really a big rake dragged by a horse, must do this work.

It took the boy longer than some of the others to do his work, for George did more work at home than the others. He was probably better informed on farm matters, however. His father was a real farmer; the other boys' fathers farmed, too, but not as a business.

Anticipating the amount of time this preparatory work would take he had not started his cabbage inside. To get an early crop of cabbage, seed must be planted in January or February; then one may start in March. But for the late crop plant in the open in May or June. This is just what George did.

He made furrows straight down his sunny southern slope. These furrows were two feet apart. The seed, of Savoy cabbage, was sprinkled in the furrows. This was done after rain. Cabbage needs much moisture for quick germination. George might have poured water into the furrows and puddled or stirred the earth a bit, if the garden had been small, but his was too large for this, so he took advantage of Nature's watering. When the plants were about two inches above ground they were thinned out to stand two feet apart in the furrow.

Cabbage, you know, is quite likely to become infested by pests. Perhaps the most common of which are lice or aphis and the cabbage worm, a green caterpillar. Therefore it is well to try a little prevention. So all over the ground about the plants sprinkle unslaked lime. Tobacco dust or soot may be used for this purpose, too. Good cultivation also helps prevent these pests.

One row of cabbage began to develop worms. These George picked off, but he found that he could not keep up with them; so The Chief advised him to buy a little pyrethrum powder at the store. This he mixed with five times its bulk of dust. Putting the mixture into an old potato sack he shook it over the infested heads of cabbage.

Except for this drawback the cabbage did well. He lost the infested row of cabbage. For he pulled them all up, spaded the ground over, and sprinkled it with the poison mixture. All the other cabbage heads were sprinkled with it, too. One may easily lose all his cabbage from these worms.

In the fall the cabbages were harvested. This was about the last of October. George pulled them up by the roots. He found some of the heads rather soft, some bursting open. As it does not pay to keep such cabbage over, these were fed to the cattle—a gift, George called it, to pay for the fertilizer.

All the fine solid heads are worth storing. In order to get nice white inner leaves, as the head begins to form break and bend over the outer leaves and those that protect the inner ones. It is a sort of blanching or bleaching process. Two hundred fine firm heads were the result of the work of this boy.

"What are you going to do with all these, I'd like to know?" asked Jack.

"I expect to store a number of them—one hundred and fifty, I should say. I'm going to give away fifty. In the winter I hope to sell about one hundred of my stored ones."

George's way of storing cabbages is a good one. A spot was ploughed in the orchard between the rows of trees. Then the cabbages were piled in a neat pile roots up, one cabbage fitting into the other. All about and over this heap a layer of straw about four inches thick was placed. To hold the pile in place stakes were driven in about its base. To hold the straw, branches were placed over the whole and boards put on last. The straw packing kept the cabbage from freezing. If George's father had had a good tight shed the cabbage could have been stored on shelves in this. The ordinary home cellar is no place for storage of cabbage.

Later in the winter he sold one hundred heads of cabbage to the markets in a near-by city. These he sold at two cents per head. They kept fifty at home.

The boys tried long and hard to find out where the other fifty went. But George would not tell. There was an orphans' home some few miles from the village. It seems that at one time an appeal had been made at the school to the boys and girls to give whatever they could to this home. At that time George had nothing to give. No one knew how badly the boy felt, so as his cabbages grew the lad made a pledge with himself to give one quarter of his cabbage to this home. One evening in late October, George had hitched up an old farm horse, loaded his cabbage in, and had driven over to the home.

The Chief learned of his kindness one December evening, when he visited the matron to see about Christmas gifts for the children. She told him that one evening in the fall a bashful lad had brought a load of cabbage to her, but would not tell his name. As the man walked home he thought of the really splendid ending of George's cabbage experiment. After all a garden reaches its real work when some of its product is given to those who are in need.

"Now I see," said The Chief out loud, as he walked past George's house on his homeward way, "why George made out of his garden so much less than the others. I never could understand why he lost the prize. I am glad there are boys who care less for money than for other things."

XII

PETER, POTATOES, AND PROFIT

Peter had a mile to go to his garden, which was on his grandfather's farm. This farm land, you will remember, was especially good.

The ploughing, fertilizing and harrowing were done for Peter. The soil was just the sort potatoes thrive on, a sandy loam. After the furrows had been made about six inches deep and two feet apart, Peter put a sprinkling of chemical fertilizer into the bottom of each furrow. This was sprinkled on as one puts salt on potato before eating it. Over this he placed some dirt so the fertilizer would not burn the potato.

Early the morning of planting Peter cut his seed potatoes. The date was the 1st of April, not a bit too soon to get in early potatoes.

The seed potatoes chosen were fair, smooth specimens of good size. These he cut so that only one eye was left to a piece of potato the size of a hen's egg. These pieces were dropped into the furrows at distances of fifteen inches apart and four inches deep. After covering, the man went over the potato patch with a harrow. A boy might use a rake for this work, but as Peter's patch was a small part of his grandfather's field the harrowing of the whole was done by the man.

When the little potato plants were well up Peter sprayed them with Paris green. This was wise because he thus got ahead of the potato bug. Some one may like to know how to mix up Paris green. The proportion used was one tablespoonful to a pail of water. This was put on with a watering pot every two weeks, thus Peter kept his potatoes quite free from bugs.

Although the rest of the potato patch was cultivated by the horse, Peter used the hoe. He could not plough, for Peter was a rather small boy for his age and not very muscular. The secret of potato culture is to cultivate well and keep the bugs down.

He dug his potatoes about the middle of June. From the one quarter acre his grandfather had lent him for his garden Peter dug seven bushels of potatoes. At the time new potatoes were selling for $1.25 per bushel. His father bought three bushels and the other four were sold in the city to Philip's mother and friends.

The constant working of the soil for potato culture gets it into a fine mellow condition exactly right for celery. Peter's grandfather suggested that the boy put this in, and so have another crop, a fall one.

Although this soil had been well fertilized in the spring for the potatoes this was yet not sufficient for celery culture. Celery ought to be started either indoors in flats, or in a hothouse or seed bed late in February—transplanted to other flats, and again finally to the open ground.

To prepare for the celery trenches were dug three feet apart and one foot wide. The earth thrown out in trench digging was piled between the ditches to be used later in banking up the celery. These trenches were six inches deep. In the bottom of the trench was put some enriched manure. This was of different materials. Peter used well-rotted barnyard dressing, a little hen manure, and about the same quantity of chemical fertilizer. Hen manure is rich, so he did not use the bulk of that. Over this was put an inch of soil.

Celery plants should be set about six inches apart in the trench. First cut them back; that is, cut off about one quarter of the root and one-third of the top. This cutting back increases the spread of root-growth later and decreases the amount of respiration of water from the leaves. The top alone grows more stocky and bulky.

Firm the plants well. That is, press the soil firmly about the roots and stalk. When the plant has received its growth it must be blanched. This process not only whitens celery, but also takes the bitter taste out of it. This may be done in various ways, but Peter used the earth process. He tied the bunches up together with bits of raffia. This was done merely to keep the earth from pressing in between the stalks. Then the earth which had been left in between the trenches was drawn up with the hoe about the stalks until only the top leaves poked out above. Do not do all this banking at once. Take several days at it. Boards may be placed along the sides of the banked celery hills.

Peter having heard of the self-blanching kinds told his grandfather that he would plant this kind to obviate blanching. But there were two drawbacks. In the first place, he had waited too late to start seed. And secondly, these varieties, too, should be bleached to take out the bitter taste. So Peter bought young celery plants from his grandfather. He paid $1 for two hundred small plants. Later in November he sold these to the same market where his potatoes had gone. Fifty bunches he sold to his father. These were left protected in the ground for use when needed. The rest he sold in the city. From his celery he made $1.80. In all the lad made $9.55. This was a pretty good sum to make. So Peter's saying of "potatoes and profit" actually worked out.

XIII

PHILIP'S BACKYARD

On a beautiful Saturday in late April one might have seen The Chief and his boys boarding the train for the city. It was the day set apart for helping Philip.

The boy had cleared up the yard ready for work. You will remember shrubs were to be planted, a walk and cement pond made, and a little gardening done. The shrubs were planted as the school trees were. One must be careful to dig the hole large enough to receive the roots of the shrub. If old shrubs are transplanted be sure to cut out all the old wood, and also cut the top back severely; that is, reduce it to one-third its former size.

It was great fun to make the little pond. Such a pond may be used for fish or for water plants or for both. Do not make too large a one, unless you have a very large yard. The smaller ones are easier to care for, and more pleasing in appearance.

First stake off the outside limits for the pond. Philip's was to be four feet by two and one-half feet. Dig down three feet. Fill in the bottom of the pond up to fifteen inches of the top with large stones. Then in between and over these put small stones so that you have filled in about six to eight inches of the cavity. Now it is time to mix cement. Mix only a little at a time. Get a board about two feet square. With a trowel put on the board one part of Portland cement to three parts of sand. Have a watering pot full of water at hand. Add water enough each time to the cement and sand to make a soft but not running mass. If it be possible for you to have small stones to put in, it will improve the mortar you are mixing. These stones should not be larger than one inch in size. Add four parts of these to the mixture.

Now over the bottom of the pond put on the paste about three inches thick. Fill in with the trowel and smooth it off with the back of this same tool. The sides are the next job. Put a board slantwise against the bottom of the pond so there is a space between the board and the side you are to plaster. Drop the mortar down into this space and press the board against the sides. This firms the mortar. Keep up this work all around the sides of the pond.

Another way to do this work is to make a box that will fit approximately into the pond, but that gives a space all around over three inches from the sides. Then the mortar may be dropped in. After three days knock out the box and you have the inside all right. If you wet the sides and bottom of the box the mortar will cling to it less.

If the mortar looks rather rough after you finish, mix cement with water, take a whisk broom and with this brush the paste all over the bottom and sides of the pond. All around the ground by the pond, mortar for about six inches. This prevents the breaking in of the edges.

Albert and George, who worked on this job, did the mortaring the first way. The pond was left unfilled for a week to dry thoroughly. Then after placing two inches of sand in the bottom it was filled with water. Philip bought two pond lily roots. He tied stones with string on the roots of the plants to keep them down: otherwise they would have bobbed up and floated on the surface of the water. Some one gave him two water hyacinths.

In the middle of the yard a round bed was made. To do this take a cord and tie a stake at either end. The cord should be whatever length you have decided shall be the radius of the circle. The radius of a circle, you remember, is the distance from the centre to the circumference.

Now drive one of the stakes into the ground at the exact centre. Grasp in your hand the other stake and swing a circle with it. The stake will scratch a well-defined line so that you have the outline of the circle, the boundary of the bed.

Jay spaded down to about six inches all along the outlines of bed. After that the bed itself was spaded. Philip insisted on outlining it with brick which had been given him. Some children use whitewashed stones, some use shells. Either plan gives a spotty effect. The idea fails of being artistic. A neat cutting of the turf and a slight heaping up of a round bed toward the centre gives after all a far more pleasing effect. Try to keep as near to Nature's own plan as you can. Shells belong on the seashore or in a collection; keep stones for road making, wall building, cement work and curbs; bricks are for foundations and buildings. Rarely use things for what they were not intended. It is better usually to border a bed with low-growing flowers. Ageratum, candytuft and dwarf nasturtiums are good for the purpose.

Along a walk to an old outhouse they planted asters on one side and four o'clocks on the other. Asters, as all boys and girls know, are better if started inside early. Then they may be transplanted to the outside. In his way one gets a bit ahead of the season.

But Philip was obliged to plant seed for both. So he planted it in a drill as one plants lettuce. Later the little seedlings were thinned out to stand six inches apart. This thinning was done when the plants were four inches high. Four o'clocks need lots of room as they grow bushy.

Plants in narrow strips are quite likely to drop over their limits. To keep these in bounds Philip later built a fence. For this he used stakes driven into the ground at intervals of every four feet. To these he nailed strips of railing. Sometimes cords are used instead of railings.

An old fence was all about the yard and an old unused outbuilding in the rear. These were both unsightly in appearance, so they had to be hidden. Vines were used for this purpose.

About six inches from the fence a furrow was made four inches deep. Climbing nasturtium seeds were dropped into this furrow at distances of every four inches. The seeds had been soaked over night. This was because the ground was very dry and the weather was now quite settled and warm. If the ground is wet and the weather cold, never soak seed. It just adds to the general soggy condition to do this.

Photograph by Mary H. Northend.

Philip's Backyard Made Beautiful by Annuals and Quick-growing Vines.

Morning-glory seeds were planted along the end of the fence by the outbuilding and all around it. After these climbing things began to grow the pretty effect of the vines was amazing.

Many times one has to train vines so they will grow where one wishes. In such cases drive small stakes into the ground back of where the vine is planted. Tie a cord or string to the stake and carry this up to where you wish the vine to go. The string may be attached in the best way, according to the place. If it is to an old building, drive a nail into the side, roof or peak of this. Some people make latticed trellises. These may be made from laths.

A neighbour gave Philip some canna bulbs which he planted in an old sieve filled with rich dirt. Canna bulbs look much like sweet potatoes. Usually a bit of stalk is left on the bulb. Leave this in planting above ground for about one-half inch. Dig a hole large enough to place the canna bulb and deep enough so the stalk comes above the ground. Place one big, fat bulb, or two or three little chaps in one spot. Leave about one foot between plantings.

In the fall after frost cut off the stalks about two inches above the ground, dig up the bulbs, shake all dirt off, and put into a box with a little thoroughly dry dirt until spring. Leave this box where it is dark and cool.

It would have been far better had Philip planted the cannas either in the round bed or against the fence as a screen. As a general rule the planting in tubs, kettles, kegs and similar receptacles is not only inartistic, but gives the plant very confined and cramped quarters. When possible plant right out in the ground. Window boxes and roof gardening in boxes is "another story."

The cost of Philip's flower garden was 25 cents. He bought five-cent packages of each of the flower seeds. The cannas cost nothing. The shrubs were $1, the cement 70 cents, and the water-lily roots 50 cents. So the total cost for changing an ugly yard to a mass of flowers was $2.46.

Philip's clearing up seemed to be catching for the girl across the street started in with her work. For ten cents she bought a collection of flower seed. These seed were planted in three-foot beds. The beds were banked up or supported by strips of board. This same girl planted flowers in two old kettles and set one upon an empty cask and the other on an old drain tile. But she later decided very wisely that this was not after all so very pretty. Kettles are better for potato boiling than for flowers.

But such a good time as she had all summer in her own green, pleasant backyard! And so had Philip, too! "Just a few cents and some hard work will change your backyard into something beautiful," Philip was heard to say one day to a group of city boys.

XIV

THE CORN CONTEST

Each boy was to take a certain number of hills of corn in his father's corn patch. He was to select his seed corn after a few suggestions given him by The Chief. These hills of selected corn were to be cared for by the boy himself, but it was perfectly legitimate that the soil be prepared for him, since most of the boys were to plant in their fathers' cornfields.

In the growing of corn the first matter for a boy or girl to consider is the selection of the seed. Corn should be selected carefully by the individual stalk; that is, choose ears from stalks bearing an ear or ears at, or a little below, the middle of the stalk.

The stalk itself should be thick and free from suckers or any evidence of disease. The ear should be cylindrical. The kernels should be deep setting, uniform and compact. Then the cob should not be too large. Look at some samples. See how some ears have too large a cob, others too small, while still others show a right amount of cob.

The butt and tip of the ear should be well filled out. Look for a perfect ear. The kernels are uniform in size, in even rows, with only a slight space between rows. See the tip and butt. Very little space is lost at the butt. You have seen ears where the butt was all space.

There is still another sort of corn. It might be called mongrel. Any one can raise such corn. Good care shows in corn as good breeding does in boys and girls.

One more point the boys were told to consider in selecting seed ears, that was the relation of the circumference of the ear to its length. An ear should have a fairly large circumference at the base and taper toward the tip. To estimate relation of circumference to length, which should be as three is to four, measure the ear one-third the distance from butt to tip. So if the ear is eight inches long the circumference should be about six inches.

All the boys but Jack and Myron tested their seed corn to be sure of its vitality. Peter went a little further than the other boys. He not only tested for general vitality, but he tested for vitality among the ears he had selected as good seed ears. This he did in the following manner: He chose twenty-five ears, and used four kernels from each ear. First a soup plate was filled with sand. This was moistened by dropping a little water on the sand. Sand must not be too wet for this work. He partitioned off the sand-bed into rows with cardboard between them. On the cardboards was marked over each row of four kernels the number of the ear from which they came. The sand was moistened each day. Peter worked out from this the best ears for quick germination.

The next point to be considered in corn culture is that of the soil. Soil for corn should be mellow and fine. If it has vegetable matter or humus in it, then its value as a corn-growing soil is increased. Fertilize well and plough, or if the plot is small, spade. You ought to have surely eight inches of good, mellow soil. In planting corn place five kernels in a hill. You will remember that a hill means a place. For corn which grows high make the hills four feet apart; while for the low growing varieties place three feet apart. Cover the early varieties with one inch of soil; the later varieties with one-half inch.

As the corn shoots begin to appear every boy should appear with his hoe or cultivator, for one secret of good gardening is constant cultivation or stirring of the soil. Water, as you know, rises in the ground and coming to the surface evaporates. Now the point is to keep the moisture in the ground for the plant's food supply. So if one keeps stirring the soil he makes a layer of earth which stops the water as it rises. We call this a mulch.

When the shoots are six inches high choose the three finest little corn seedlings in the group of five. Pull the others out. The reason for putting in five kernels in the first place, instead of three, is that some may not come up. And, too, some that do come up may be poor and sickly.

Myron did a very stupid thing. At least he called it stupid. Some one sent him a packet of seed popcorn. Myron thought it would be pretty interesting to raise some and supply the club with popcorn at its meetings all the next winter. Now Myron did not know that from the corn tassels the pollen when ripe or dry blows all over the corn field. This pollen falls on the silk of corn plants anywhere in the field. The pollen fertilizes the plant and the ear of corn sets and grows. Because the pollen being light is blown to such distances and because different kinds of corn can interpollinate, is reason enough for not planting different varieties of corn in one patch.

Myron's popcorn and sweet corn fertilized each other and he got a corn which was a cross between the sweet corn and popcorn. He learned a lesson of pollination, but at the expense of the corn crop.

One may plant early and late corn in the same patch but otherwise he should stick to one kind of corn.

The boys in the fall were to submit twelve of the finest ears they had raised. These were to be scored or sized up as follows:

NAME OF CONTESTANT	VARIETY OF CORN	DATE OF PLANTING	
(1) Ear: A. Trueness to type B. Shape		10 10	20
(2) Filling of: A. Tip B. Butt		10 10	20
(3) Kernels: A. Shape B. Arrangement of cob C. Depth D. Colour		6 12 6 6	30
(4) Measurements of ear: A. Length B. Circumference		10 10	20
(5) Proportion of Corn to Cob:		10	10
		Total	100

This score card needs a little explanation. Take up (1) Ear, first. All the twelve ears presented ought to be much alike; that is, like the type or parent ear you are striving to produce again. So if, out of twelve specimens, six were fine ears and the other six were rather poor, then surely ten credits or points could not be given. The shape of an ear should in general be tapering, well rounded a little below the centre, and tapering not too abruptly toward the tip.

The second point is the Filling of the Tip and Butt. The tip should be filled with even, regularly arranged kernels. It should not be too pointed nor too blunt. The butts should be covered over with kernels except where a deep, clean-cut depression is left. Here, as in the tips, the shape has to be considered, for flattened and blunted butts are bad form.

As to Kernels, they should be uniform and well-shaped, not only on the single cob, but in all the specimens. The furrows must be uniform, regular, and with only a slight space between. To determine depth cut a square inch right out of an ear. All kernels should be of one colour. If a red kernel is in with the white then there has been an intermixing. All kernels on all ears ought to have the colour of those of the type form.

Not only should there be a proper proportion between Length of Ear and Circumference, but an ear may be too long. The usual length of ears is from eight to twelve inches, according to variety.

We have spoken of cob-relation before. This cannot be determined accurately by your eye but must be done by weight; so shell the corn, weighing the ear first. Now weigh the cob. The difference is weight of corn. Divide the weight of the corn by the weight of the ear. This gives per cent. of corn. For the exhibit the boys afterward used half their samples submitted and reckoned per cent. on this. The proper percentage of corn to cob should be 86 or 87 per cent. You can easily find out if you come up to standard.

Myron brought in some corn merely to show his mistake but of course did not submit any.

Jack, you will remember, did not test his corn and results showed this. Out of his twelve samples there were two good ears. The others showed many changes. The poorly filled tips, irregular rows, and wide space between rows—all these scored against Jack. George's corn was thrown out because black kernels were found here and there in with the others.

Albert's and Jay's Peep-o-day came out in fine shape. But Peter's Country Gentleman after all had the record. Philip dropped out of the race because he went on a summer vacation. So for a slight amount Peter took over Philip's corn hills.

That fall the boys made very careful selection of seed corn.

"After all," Myron said one night at club meeting, "although Peter's corn was the only really fine specimen, I think some of the rest of us got fully as much out of the corn contest."

"So do I," added Jack; "and I, for one, shall test corn after this."

"I think our corn was pretty good," Albert went on in a half-injured way; "but we are going to beat that record next year. We shall rotate our crop, planting our corn where the beans were this season. That's a thing fellows ought to know; that it's a mighty good thing to rotate crops."

"What's that?" asked Philip.

"Rotating crops means not always planting the same crop on the same piece of land, but changing every two or three years. It happens that beans are very good to plant before corn. They do not take from the soil, The Chief says, what the corn needs. So a piece of soil planted to beans gets in shape for corn planting another year. It would not be well to plant corn on a certain piece of land more than two successive years. Then something else should be planted on this land and the corn put somewhere else."

"Good!" said The Chief. "Some day we shall discuss rotation of crops more fully. There are no end of topics for us to work over this next winter."

XV

THE GIRLS' SECRET WORK

The girls were each to raise something special at their own home and then each was to have a share in a big garden. Katharine, who had quite a lovely yard, was to give the space for the general garden. This was largely because Katharine's home was on the river road, a bit out of the village and near none of the boys' places, for the girls wished to keep the knowledge of their work from the boys as long as possible. Helena lived next house to George and the land she might have used for this big garden would have bordered George's corn patch. So that, of course, would never do.

The garden while formal was ornamental. The girls were all to work on the staking out and preparation. But each girl was then to take a section of it and plant and care for that.

Katharine was to take the centre portion of grass and cannas. Now a grass plot is very pleasing in a garden. It is restful to the eye and is much more harmonious with the other colours in a garden than a mass of brilliant blossoms. Cannas have some height, a delicate splash of colour in the blossom and so work in well. It is always well to put some tall-growing plant in the centre. The effect is that of working up to a climax. One should not immediately jump from very low flowers in the beds to a few tall ones in the centre. This is ludicrous. Make the gradation gradual from low to high.

This garden of the girls may seem almost to violate this principle. Not so, for the nasturtiums merely acted as a border. Then all around the garden were the zinnias, poppies and marigolds a step up to the cannas. One may buy tall or rather low growing cannas. These latter grow about four feet high. They chose these low ones with yellow and orange in the blossom to harmonize with the yellow and orange of the nasturtiums.

Note the proportional amount of grass space in the girls' garden. Observe too that it is the centre of interest.

The nasturtium border was Elizabeth's. Zinnias were chosen by Helena, and Katharine was to help in this work. Eloise loving the mignonette had asked for it, poppies were Josephine's and marigold was for Dee. Ethel wanted the border of sweet alyssum although it represented a long strip to work.

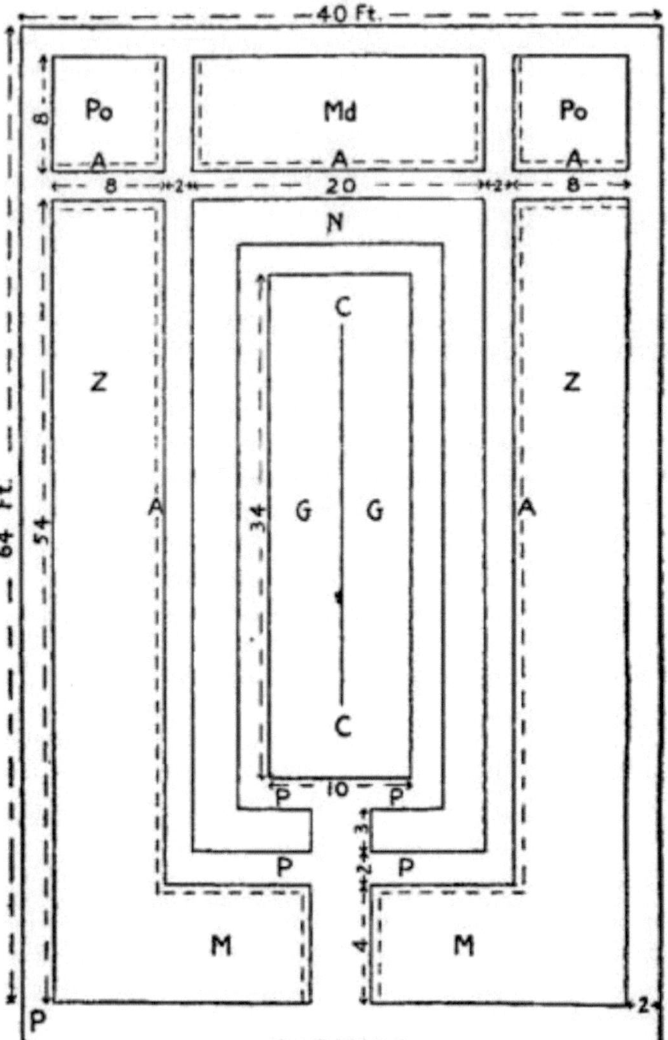

C = CANNA

M = MIGNONETTE
Z = ZINNIA RED
N = NASTURTIUM
G = GRASS

A = SWEET ALYSSUM
Po = POPPY RED
Md = MARIGOLD
P = PATH

Diagram: Scale ⅛ in.-1 ft. Note the proportional amount of grass space in the girls' garden. Observe, too, that it is the centre of interest.

If you think over this garden with its brilliant poppy colour, the heavy yellow of the marigold, the lighter colour effect of yellow in the nasturtium, the dark red zinnia—quite a splash of colour, was it not? In order to have great masses of brilliant colour in the same garden one must break them in some way. There are two possibilities that are good: first, paths between beds, and second, borders of white or inconspicuously coloured plants. Sweet alyssum is good for this purpose and so too is mignonette. Mignonette has such a small and modest little flower that one thinks always of mignonette in terms of green. The mignonette was massed at the entrance of the garden for pleasing and subdued effect.

In staking out a garden it is well first to put heavy stakes, like the ones the boys made, in each of the four corners of the entire plot and put a string around. This strings off the outline of the entire garden. Sight along the lines as Jay and Albert did to be sure the lines are straight. In sloping land true up with great care. On a level bit of land sighting is easy.

Next if there is a main path string that off using twine, stakes and always sighting. If the garden is a large one of many small plots the next thing to do is to string lengthwise the entire garden, measuring off plots and paths. Then widthwise measure off paths. The side paths give you the width-boundary of the plots. But the paths have cords through portions of them. So put stakes in the corners of the plots. Cut the cord in the centres of cross paths and tie to stakes. If carefully done each plot is marked off shipshape.

In general make the main paths four feet wide, plot paths or side paths eighteen inches. Plots with widths about one-third of the lengths are right. This did not hold true for this garden since the beds were long narrow strips. In such cases the width should be a comfortable one to lean over and work across.

In staking such a garden plot as the girls' it is well to first line off with stakes and cord the entire outline of the garden. Then next string off main paths and intermediate ones. It is very easy then to string off the beds, for the path boundaries have done most of the work for you.

The girls planted all their seeds except poppy and grass seed in drills just as one plants lettuce and radish. This is a far easier way to plant since as the little seedlings come up one can easily distinguish the nice even row of little plants from weeds. They decided later that it would have been easier if the poppy had come up in drills. For it came up in little tufts here and there. And, sad to say, the poppy does not stand transplanting.

In making drills take two stakes and a string. To either end of the string tie a stake. The length of the string after tying should be the length of the drill one wishes to make. This will be usually either the length or width of the bed. This, then, is your line for the drill making.

Another thing the girls did which makes garden measurements easy is the following: mark off on the rake or hoe handle three feet. One of these feet measure off into inches. This saves carrying a measuring rod into the garden. The marking should be done on the edge of the handle and not on either the under or upper surface of it. If garden stakes are made one foot in length they can be used in measuring.

Nasturtium culture needs some explanation. Nasturtiums are the most accommodating of flowers. They will live on almost any soil. The seeds are large and so are very easy for little girls to handle. They may be placed two seeds together six inches apart in the furrow. If the soil is very dry and the weather very warm, soak the seeds over night. Plant the seeds about one inch deep, cover over the soil and firm it well. It is easy enough to keep the nasturtium bed weeded for the seedlings are large and not to be mistaken. Keep the flowers well picked all summer and you will have numberless blooms.

Sweet alyssum is a charming border plant. This, too, grows in almost any soil. It is well to sow the seed in a box indoors. Transplant when the little seedlings are two inches high. But alyssum may be sown right outdoors in the garden plot. Sprinkle the seeds along in the drill. After the seedlings come up and are about an inch high thin out until the seedlings stand six to twelve inches apart.

Marigolds are very gay sort of flowers. Many do not like them on account of their disagreeable odour. But a strong point about these flowers is this: they bloom and bloom, and then they bloom again. There are three kinds of marigolds one might plant. These are the African, French and dwarf. They differ in height and also bushiness. The African varieties must be thinned out to stand fifteen inches apart, the French ten inches and the dwarf six inches. These seeds are dry, dead looking chaps, but are not so small that they cannot be handled separately and placed carefully in the drill. Plant them nearer together than they are to stand later. For instance, put the African five inches apart, the French five inches, too, and the dwarf three inches. Then you have extras, so if some do not come up your garden is not crippled.

Mignonette, again, is accommodating and will grow in almost any kind of soil. These seeds are small and may be sprinkled along in the drill. Later thin out so the plants stand from six to twelve inches apart. In choosing mignonette seed remember that there is a great difference in mignonette. Some is very sweet, some is not: some have large sturdy spikes, while others have rather small spikes. It pays to buy good seed.

Poppy is a trifle more particular about the soil it grows in. It requires a rather rich, sandy loam. Again remember that poppies never stand transplanting. Poppy may be planted broadcast or in drills. The tall growing varieties should finally stand eighteen inches apart and the smaller ones nine inches. In order that poppies may blossom freely you should never let a seed capsule form. For you see that if one wishes bloom, one must not let the strength of a plant go to any other work except blossom making.

Zinnias are satisfactory just as marigolds are. To be sure they are not a very graceful flower. But what of that? We need all kinds of flowers. When you buy the little packets of seed you usually get a mixture of colour.

In order to have just the colour one wishes, seed must be bought from the seedsman by the ounce. The girls wanted dark red zinnia. One ounce planted this space. It is not as expensive to buy seed this way. A number of people may club together for seed.

Helena's method of planting zinnia was to sow the seeds in a drill. Later she thinned her seedlings so that they stood eight inches apart in the row.

Cornflowers or bachelor's buttons are lovely too. They are far lovelier if bought by bulk so one may have the one colour, that lovely blue. These seeds may be planted in drills two seeds every six inches apart. Later thin to twelve inches apart.

Most people start cannas from the bulb. When one does this, plant a good sized bulb and leave about an inch of stalk above ground. If the bulbs are smaller use two to a hole or planting. If cannas are started from seed follow this direction: File holes in the canna seed. The reason for this is that the outer crust is tough and filing helps the young plant to get out. These seeds should be soaked in warm water for a day. Plant in pots. When the plants are six to eight inches tall transplant to the garden. Cannas should stand two to six feet apart. It depends on variety, whether tall or dwarfed, how far apart to place them.

When the flower garden was first started the question arose: "Shall we plant annuals, biennials or perennials."

"For my part," said Josephine, "I don't know at all what these words mean."

Katharine got a dictionary and soon she and Eloise had these botanical terms worked out as follows:

A perennial is a plant which lives year after year in the soil. It usually blossoms its second season. Trees and shrubs are hardy perennials.

A biennial is sown one year, blossoms the next and then dies. Biennials should be covered lightly with straw or leaves through the winter.

An annual blossoms and dies its first season. But some annuals sow themselves and so come up again the next season.

The girls worked out a table of planting by months which Ethel called the plant time-table.

Besides the garden which the girls all had together each one did something to improve things at home.

THE FLOWER TIME-TABLE

NAME	SOWING TIME	BLOSSOMING TIME
Ageratum	May	June-October
Aster	May	Until frost
Balsam	May	June-September
Calendula	May	June-October
Cal. Poppy	May	August
Candytuft	May	June-September
Coreopsis	May	June-August
Cornflower	April	June
Cosmos	May	August-September
Four o'clock	May	July-August
Foxglove	May	June
Gaillardia	May	July-October
Helianthus	May	August-September
Hollyhocks	August	August
Iceland Poppy	May	June-September
Larkspur	May	June-July
Marigold	May	Until frost
Mignonette	May	Until frost
Morning-glory	May	July-August
Petunia	May	July-September
Phlox	May	July-October
Scabiosa	May	July-August
Stock	May	June-July
Sunflower	April	July-September

Ethel and Dee set up a sundial in Ethel's own backyard. The directions that follow will help other girls and boys in setting up theirs. Sun-time and clock-time are not quite the same. There are four days in the year when, if you work out the sun-shadow time, your dial will be almost accurate. This is because on these days the sun-time and the clock-time practically coincide. These dates are April 15th, June 15th, September 1st, and December 24th.

Before you go outdoors draw on the platform of the sundial a straight line from angle B of the gnomon to the front edge of the platform.

Set the dial out in direct sunlight. The shadow cast must fall right on the straight line which you previously drew. When the shadow and the line coincide, mark the extreme end of the line XII. This stands for twelve o'clock. Now screw the sundial in this position to the column you have made for it to rest upon. At one o'clock mark where the shadow points, and keep on with this for every hour.

You remember the gnomon angle was the number of degrees corresponding to the degrees in latitude of your special place. Poughkeepsie boys and girls will be interested to know that if a sundial be brought to them from Rome, it will be right for them. And if New York City boys and girls could get one from Florence, they would find it accurate for their own use. These girls lived near Poughkeepsie.

Elizabeth planted a border of nasturtium, sunflower and zinnia along her sidewalk. It cost eight cents for seed to plant these two by ten feet strips.

Helena made a bed of different kinds of flowers right back of her father's field. The garden was thirteen and one-half feet square. The edges her father helped her sod, this making a terrace effect. Nine little flower beds were marked off with paths between. In the beds were asters, celosia, balsam, nasturtiums, marigold, zinnia, carnation, schizanthus, sweet peas, dahlias, gladiolus, candytuft, lilies, scabiosa, stocks, salvia, snapdragon, phlox, mignonette, four o'clocks and petunias. Helena's mother worked with her in the garden as did one of the boys across the street. He was not a club member but was hoping to be one the next year. And so Leston worked with Helena all summer long. He finally won his place in The Chief's club.

Eloise decided she would have a window garden and so before all the front windows of the inn, window boxes were placed.

Most of the trouble with the window box is a lack of drainage space. Estimate off the bottom of the box something like this: To every foot bore six holes. This is none too much. The great trouble usually is lack of drainage, or lack of air, or sour soil. Over each drainage hole put a bit of broken pot. Then it is well to put a half-inch of drainage material in the box. Stone, broken pot, sphagnum moss, or hay will do for this. The soil should be good, rich, garden soil. With this one might mix in some sand to help drainage. Window boxes should be watered with care; they should not be flooded.

Eloise had very effective boxes. Vincas trailed over the edges; dwarf cannas were in the back of each box; and red and white geraniums were a glory all summer long.

Josephine's gardening was a little difficult. She had no space at all. The backyard at her house was seeded down and her mother did not wish it spaded up. She had no front yard. Josephine thought and thought for some time, then decided she would just simply have to make a way to have a garden.

So one day she went to the grocery store and bought a soap box for ten cents. This she filled with soil from Eloise's garden. Then she bought a five-cent package of parsley seed. These seeds were soaked over night in warm water, for parsley seeds are slow to germinate.

Then the seeds were planted in neat little rows in her box garden. This garden was most convenient. It stood out near the house in the backyard all summer. It went to the exhibit in the fall. It stayed on the piazza until frost and then went into the kitchen for the winter. Josephine had parsley enough for her mother's table all the year around.

XVI

MORE ABOUT THE GIRLS' WORK.

In late September the girls began agitating the matter of bulb planting for the school grounds and their homes. The boys were rather scornful of it.

"I believe in gardens," said Albert with great finality, "but bulb work seems to me like fancy work. And then too, bulbs are pretty expensive."

"Very well," answered Dee, "we girls are quite able, as you boys know, to work alone. But spading is pretty hard, and I should think some of you would be glad to help."

"I'll help any time," Myron volunteered, "and I promise to bring two of these other chaps whenever you say."

"Thank you, Myron. We'll not bother you boys further now." Off the girls ran to Katharine's home to study bulb catalogues. Katharine's father gave five dollars for bulbs for the school grounds. This he stipulated was for outdoor planting. Elizabeth and Ethel were going to plant outdoors at home. The other girls had each some money for indoor work.

You may all like to know what the girls found out from their search in bulb catalogues. In the first place very good and perfectly reliable information is obtained from the catalogue of any reputable seed house. The girls found out that certain bulbs are better adapted to outside planting, while others do equally well indoors or out. Take tulips first; these are suited to the outdoor conditions. To be sure the florist, whose business it is to raise them inside does so with great success. But boys and girls are more likely to have trouble with inside planting of tulips than of other bulbs. Oftentimes lice cover them when the bulb is first brought up from the cellar. Then when treated with kerosene emulsion or some other insecticide the bud becomes blasted, for the blossom is close under the folded outer leaves, so is in a very precarious position. Then, too, tulip bulbs rot easily and the buds blast easily. So it is wise not to run so many risks but try the kinds of bulbs which are less prone to trouble. The easiest and safest bulbs for children to work with are narcissus (including daffodils, jonquils, Chinese lily bulbs and paper narcissus), and hyacinth.

Hyacinth has one bad habit when planted indoors. This is the tendency to unfold its blossom too soon. So the beautiful hyacinth blossom appears dwarfed and stunted close down near the ground. To avoid this condition do not take the bulb from the dark until the leaves are about an inch to two inches above the earth and until they have spread apart. This gives the blossom a chance to shoot up. Tip the pot over and see if the roots are visible through the drainage hole.

The time to buy bulbs is in late August or early September. After this bulbs through shrinkage depreciate in value; by which value is meant not one in price but in soundness and ability to produce blossoms. Do not buy cheap or cut-rate bulbs. Buy good, big, sound ones.

The Roman hyacinths are excellent for forcing. They are small flowered, quite different from the large sturdy Dutch hyacinths more commonly planted. In choosing hyacinths you have to decide upon the colour and whether you wish double or single varieties. In general most people enjoy single flowers better. If you are to use the hyacinths for outdoor planting or bedding it is perfectly safe just to write for bulbs which are to be bedded. La Grandesse is a beautiful white; King of the Blues speaks for itself and the Sarah Bernhardt is a salmon pink. These do well inside, too. Charles Dickens is a fine rose colour, Prince of Wales, violet, and L'Innocence, a fine white. These are good for inside planting. Some may like the smaller Roman hyacinths, which do splendidly indoors. Very good hyacinths are bought for fifteen cents.

Tulips do especially well outdoors. A capital one for either bedding or indoor forcing is the Isabelle. It is a beautiful red tulip which is bought for five cents. The Summer Beauty, a hardy white tulip, is well worth the ten cents asked for each one. Some of you may like to raise some freaks; then try parrot tulips at about thirty-five cents a dozen. A thing to remember about the indoor planting of tulips is this—tulips, more than other bulbs, are likely to have plant lice, so watch out!

In daffodils you may be sure of the Van Sion. These are worth forty cents a dozen. You can buy daffodils for twenty.

If you wish to lay in a stock of bulbs for water planting choose, of course, Chinese lilies, but try, too, the paper white narcissus. These bulbs cost forty cents a dozen. Buy from the five-and-ten-cent store a glass dish, and gather stones for it. About three weeks before you wish blossoms plant a dozen of these narcissi in the glass dish with the stones as a foundation, and water enough to come up around the base of the bulbs. It is a good plan to set the dish of bulbs in the dark for four or five days.

You can grow hyacinths in water too. For this a special glass is sold, although I have seen children place a bulb in the top of a preserve jar. It works all right. Bulbs must never drop low into water or they decay. These, too, should be placed in the dark for about a week.

Suppose you have a quarter to spend. You can make all kinds of interesting combinations. Three daffodils for ten cents, a hyacinth for ten, and a tulip for five, give you a chance to experiment.

A word more about narcissus. This is a large family, One gets confused sometimes with the names daffodil, jonquil and paper white narcissus. All these are of the family narcissus. The daffodils are the bulbs with large single or double cups. The jonquil has a cluster of small blossoms of from three to six single flowers. The paper white narcissus has four to twelve single blossoms to the flower head.

Ethel and Dee had good lawns at home which their mothers were not willing to have spaded up, but they gave consent to the girls putting crocus bulbs here and there over the lawns. These bulbs should be planted about an inch deep and three inches apart in the group. These were dotted about in clusters of six. The dibble is a good instrument to use in dotting bulbs around the turf. Crocuses are good for indoor planting as well. They may be planted in flats or in indoor boxes. Remember crocuses are of practically no use for cut-flower purposes.

The school tulip bed was made just as Philip's round bed was made. The time to plant depends on the weather. It is always well to get all outdoor planting done before the time of frost. Why? Because you wish to get the bulbs in while the earth is still warm. Bulbs lie in the ground all winter slowly putting out roots, slowly starting to push up toward the light above. For good root forming they need this long time of slow growth. Get the outdoor bulbs in the ground during September.

Before this the ground may be prepared. In all the beds dig down about two feet. Work over the soil well. Make it fine and free from lumps and stones. Ordinary garden soil will be right for these beds. Put no fertilizer in. If your ground is clayey, mix sand with it. Use one-quarter sand in a mixture of this sort. This makes a lighter soil. Clay soil is what we call a heavy soil. Bulbs require light soils.

Now comes planting. Different kinds of bulbs require different depths of soil. Place the tulip bulbs four inches deep, and six inches apart.

Hyacinths were planted by Elizabeth in a strip beside the house. Jack also planted daffodils in a corner sheltered by the house foundation and an old high fence. The daffodils were planted exactly as the tulips, but the hyacinths were planted six inches deep, instead of four.

In buying bulbs for outdoor planting ask for bedding bulbs, while for indoor work buy forcing varieties.

One bright day in October the girls met at Katharine's house to pot their bulbs for winter. Some had made wooden boxes or flats during the winter; others had bought low pots; while still others had the ordinary high pot.

In potting bulbs or any other plant two things are to be kept in mind—first, the soil, and second, the drainage. The soil may be any good garden soil. To a given quantity one may add one-fourth rotted manure and one-fourth sand. This last helps lighten the material, allowing more air to get at the entire mass and making good drainage easier. Mix all this together. If one lacks the well-rotted manure and sand, any good garden soil may be used. Sift the soil until it is perfectly fine.

A simple sieve, which works well, may be made from a small soap or starch box. Knock the bottom out and use in place of this wire netting. Helena and Eloise made two sieves which did for all the girls. Eloise also made some very good flats as described before under the chapter on the girls' winter work. You can easily see how excellent this style of flat is from a drainage point of view.

More trouble, in potted bulbs and all kinds of plants, comes from too little drainage space than from any other one thing. Most boys and girls think it enough if one little stone or piece of pot is put in the hole of the flower pot. Not so; there should be from one to two inches of drainage material in the pot. That seems a great deal, doesn't it? But it will give not only drainage but air space, too, and this keeps the plant in good healthy shape. With too little drainage area the earth in a pot gets clogged and very often sour. A high pot needs more drainage matter in it than a low one. First use a piece of broken pot to place over the drainage hole. But put this in such a position that the drainage hole will be kept open. Then put in two inches of coarse material like broken pot. It is now a good plan to place over this a layer of coarse material. This gives a greater opportunity for air. Over this goes the soil you have already prepared. Place bulbs just below the surface and have soil one inch below the top of pot. Narcissus and hyacinths may be planted with their tops out of the soil.

A low pot needs less drainage material. Some pots have sphagnum moss over the drainage. Instead of this use old sod finely torn up or coarse soil. See, too, that the bulb comes nearly to the top of the soil. When indoor bulbs are planted at some distance below the surface of the soil they have too much work to do to force their way up and out. It takes too long.

After the girls had finished potting the next step was to make arrangements for the resting time. Bulbs should stay in the dark and cold from five to ten weeks. It is difficult to give an exact time as conditions differ and bulbs too.

Bulbs may take their retirement in a dark cold cellar where there is no danger from mice. Some attics are suited for this. Eloise put hers in an old bureau. This bureau was in an unused, cold room. The bulbs were placed in the drawers which were then closed, but not tightly. Ethel, Dee and Josephine put theirs in the cellar. Helena, Elizabeth and Katharine tried another plan. They had a trench dug outdoors two feet deep and eighteen inches wide. In this they placed their pots and flats. Then the trench was filled in with dirt and over this a layer of ashes was put. The pots were given a good watering before they were sunk into the ground. Unless the winter is a very dry and open one the bulbs will need no more water. If there should be little snow-fall then it may be necessary to water the ground where the bulbs are, but this is not usual. Little sticks were put into the ground just where the bulbs were. These help in locating them when digging-up time comes. The girls left them in the ground for six weeks. Then they were taken in and put in a cold north window for a week. Helena put hers in the dark a week and then brought them to a north window for another week. Then she put them in a south window.

Bulbs should go very slowly from dark and cold to warmth and light. This is a point to be remembered.

The girls who stored their pots in attic and cellar of course had to water them. This should be done as often as the plant needs it, perhaps three times a week.

THE BULB STORY

1. Ample Drainage
2. Depth of Planting
3. Perfect Root Development
4. Ready to Come to the Light

Photographs by H.E. Angell and Henry Schultz.

When the plant is about an inch above the ground, as one of the pictures shows, it is the time to bring it to the light. Be sure the outer leaves have spread apart in the ease of hyacinths and tulips; also invert the pot and see if the roots are poking through; this is another way to be sure that the bulb is ready to come to the light.

A way to help a hyacinth or tulip develop its stems properly and so prevent blossoming low down in the box is to put a cone over the bulb as soon as you bring it to the light. Make the cone of paper and have one opening at the top two inches in diameter. The flower stem and leaves of the bulb will quickly start to grow up to the light. Take this cone off when the leaves are several inches high.

The girls did some water-planting, too. For this purpose hyacinths, Chinese lilies, paper narcissus and jonquils are good. Some people put these dishes and glasses immediately in the light. But it is better if they are set away in the dark until the shoots start and the roots, too, begin development. The girls bought glass dishes at the five-and-ten-cent store. Into these dishes were put small stones which they had gathered in the fall for this purpose. Stones should be small for this work, from one-half inch to an inch in diameter. Josephine had a lot of fine white sand which she packed in all about the stones. The sand was kept thoroughly wet all the time. This is a good method of treatment. Paper white narcissus, if planted early, will blossom by Thanksgiving. They may be held back until Christmas. These blossoms are sweet smelling and very graceful in appearance. Eloise tried the same method with jonquils with excellent results.

In February they had a bulb exhibit and their display was really fine. In the spring they all felt that the outdoor work, too, had paid. The beds were uncovered as early as possible. The outdoor bulb will stand considerable cold, even after it is well up. Cut worms may trouble the hyacinths; if so try the paper collar treatment.

Finally remember then three things about bulbs: good bulbs, good drainage, plenty of time in the dark.

After bulbs are through blooming let the blossom dry thoroughly and the leaves get yellow and dry. One need not keep these homely looking plants in the living room in plain sight. Put them away down cellar to finish drying out. Then cut the leaves and blossoms off to one inch of the bulb itself. Take bulb out of pot, shake all dirt off, and dry thoroughly. These may be put into paper bags and hung up in a dry place or just packed in a pasteboard box. These indoor bulbs may be planted outdoors in the fall. They will do better for the change. Indoor bulb culture means forcing, a hard strain and demand on a bulb. Outdoor planting gives a long winter's rest, not entire rest, to be sure, but the growth during the winter is slow.

LIST OF BULBS BY COLOURS

WHITE FLOWERS

NAME	DEPTH TO PLANT	OUTDOOR BLOOM	INDOOR BLOOM	HEIGHT	SPECIAL POINTS
Snowdrop	2 in.	March	Christmas	3-4 in.	Earliest of spring flowers. Likes cool and shady spots
Crocus	2 in.	March	Christmas	3-5 in.	Buy *Crocus biflorus*. var. *argenteus*
White crowned daffodils	3-4 in.	April	January	10 in.	Increases very rapidly
Poet's narcissus	3-4 in.	May	January	12 in.	Excellent for outdoor work
Grape hyacinth	3 in.	March	January	4-6 in.	Plant in shady places in the lawn
Bell-flowered scilla	3 in.	May	Christmas	12 in.	Poet's narcissus and this scilla planted together bloom at same time
Star of Bethlehem	3 in.	May		12 in.	Most satisfactory for bloom
Tulips	4 in.	April	January	6-12 in.	Try Duc Van Thol and d'Immaculée

Hyacinths	5 in.	April	January	6-12 in.	Plant Baroness Van Thuyl, very large and very early

YELLOW FLOWERS					
NAME	**DEPTH TO PLANT**	**OUTDOOR BLOOM**	**INDOOR BLOOM**	**HEIGHT**	**SPECIAL POINTS**
Crocus	2-3 in.	March	Christmas	3-5 in.	Plant Cloth of Gold
Trumpet daffodils	4 in.	April	January	15 in.	Van Sion is especially satisfactory. Flowers double
Jonquil	4 in.	April	January	12 in.	Flowers sweetly fragrant
Daffodil	3 in.	April	January	4 in.	Variety *Narcissus Bulbocodium*
Tulips	4 in.	April	January	12 in.	Due Van Thol, yellow
Hyacinths	5 in.	April	January	12 in.	Choose from those called bedding varieties

BLUE FLOWERS					
NAME	**DEPTH TO PLANT**	**OUTDOOR BLOOM**	**INDOOR BLOOM**	**HEIGHT**	**SPECIAL POINTS**
Common	2-3 in.	March	Christmas	5 in.	Good for

crocus					lawn effects
Grape Hyacinths	2-3 in.	April	January	15 in.	Good for pot culture
Iris (Spanish)	4 in.	May	January	12-24 in.	Plant in groups for garden effect
Iris (English)	3 in.	June	January	12-24 in.	If planted outdoors in September, cover well with leaves
Bulb-flowered scilla	2-3 in.	May	January	12 in.	Satisfactory for bloom
Hyacinths	5 in.	May	January	12 in.	King of the Blues

RED FLOWERS

NAME	DEPTH TO PLANT	OUTDOOR BLOOM	INDOOR BLOOM	HEIGHT	SPECIAL POINTS
Peony	4 in.	May		2-2½ ft.	Plant outdoors in September. Increases and lives a long time
Late peony	6 in.	July		3 ft.	Showy and attractive
Red speciosum lily	6-8 in.	August		3 ft.	Protect through the winter, leaving bulbs in the ground
Tulips	4	April	January	6-18 in.	Isabelle Choose vari-

XVII

THE GIRLS' WINTER WORK.

"We want some plants at school this winter, and we each should like some plants of our own at home." This remark greeted The Chief one day in late September as he entered his home after a long tramp in the woods.

The slant rays of the late afternoon sun and the low fire in the fireplace were not able to give The Chief any clue as to the speakers. "Who are '*we*'?" he demanded.

"I am Dee," was the reply, "and 'we' are all the girls."

"Dear me" said the man, "I thought I had settled your case by recommending bulb culture to you."

"Not much!" shouted the girls all together. "We have finished our bulb work," Katharine went on to say, "and now we are very anxious to do something with house plants. We have a good six weeks or more to wait for our bulbs, and so we thought possibly you would be willing to help us."

"I did think," grumbled the man, "that after I had invited you to a series of talks this winter you would leave me in peace."

And then they all laughed gaily together.

"Well, what is your stock you have to work with, girls? I shall have to know that before I can help you."

"We have—that is, most of us have—a lot of old straggly geraniums in our gardens. Then Katharine's mother has some fuchsias and begonias which she has promised us," replied Miriam.

"Up at the hotel where Jack sold his lettuce there are a few things I have been promised," added Elizabeth.

"Do you know what these are?" asked Ethel.

"Yes. There are some heliotrope plants, marguerites, some lovely rose geraniums, and a few flowering maples or—I have forgotten the long name for them."

"Abutilon is the other name," added The Chief. "Well, that is a start, surely. I'll do some potting with you next Saturday afternoon. That will give Elizabeth time to get her hotel plants. I guess Dee will drive you up. You are to take a big basket with you, and your trowels. Carefully lift each plant from its resting-place. Water the soil a bit before you take up the plants. They come up easier for this, and soil is more likely to remain clinging to the roots. If it should rain Friday you will be saved the trouble of taking a watering pot with you. Be sure to take up with the plant some of its own soil. Then pack all these soil-encased plants in your basket. Do not let the sun get at them before we get at potting. Come all of you at two in the afternoon. Bring your plants with their own earth, your straggly geraniums, pots, and each a trowel. Now perhaps you will be willing to trot home so I may eat my supper."

Next Saturday at two a grand collection of girls, plants, big pots, little pots, and trowels arrived. The Chief took girls and all out into his potting shed. This was once an old woodhouse; now a shed with benches running along two sides of it. Under the benches were great heaps of soil. Pots and pans were piled in one corner and garden implements were neatly put up on the walls.

"I call this a pretty nice place for work," said Eloise in tones full of real interest. The Chief nodded smilingly at her, for there was a bond of sympathy between the man and this real outdoor girl. Eloise had a greater appreciation of the work than any one of the others.

"Where did that splendid window box come from?" asked Josephine.

"That is one the boys made last winter especially for the school. I shall have to give you girls some group work first. Then I'll demonstrate potting and slipping to you all together. Eloise and Josephine will start to put the drainage material into the pots. Ethel and Dee may do the same for the window box. Put in your curved pieces of pot over the drainage hole, then about an inch of drainage material. There is a wooden mallet. Crack up some bits of old flower pot as you need them. Outside is a half barrel of old pots. Instead of using all pot for this half inch of drainage material, use some charcoal. In that barrel marked charcoal you will find plenty of pieces. The charcoal is not only good for drainage but helps keep the soil sweet. Helena, Miriam and Katharine will mix the soil. Here are some firkins and peck measures. To every three measures of soil from that pile there, which is nothing but garden soil, add one measure of sand and one of leaf mould. Now, my leaf mould over there in that tub isn't real mould from the woods. You see the part desired in leaf mould is vegetable matter. I can get that from old rotted leaves and rotted sods. Notice, girls, that you see no green grass in that soddy matter I have shaved off with my spade—only the under surface of the sods. This surface is full of vegetable matter in the form of young roots. Stir up all these portions thoroughly.

"Now, Elizabeth, look at these pots. Some are brand new, some are clogged with soil and green matter. Soak the new ones in a pailful of water and clean and wash the dirty ones."

"I'd like to ask why I am to soak the new pots, and why, since soil is going right back into these old pots, I have to clean them. I should think the soil clinging to the sides would help out."

"Yes, I should like to know that, too," said Miriam, stopping her work.

"If new pots are not soaked the soil in them dries out very rapidly. You can see that would be bad. Old soil clinging to plants interferes with the new root growth while the green affects the porosity of the pot."

The girls stirred, scraped, and placed bits of pot in boxes and pots for a time without speaking.

"Are we putting the right amount of drainage into these pots?" finally questioned Eloise.

"Not quite enough into those large pots. In the small ones the inch of drainage you have in the pots is quite sufficient, but in those pots over six inches in size put an inch and a half of drainage material. There should be two inches of drainage in that window box. Katherine, you are taking those new pots out of the water too soon. Leave them in until the bubbling is all over. Wait a minute, you mixing girls over there. You mustn't put such coarse soil into your mixture. That could well go into the pots and window box above the drainage material. But it is far too coarse for a good potting soil. What soil you need for potting one of you should sift. If the lumps were not so large it would have been all right. What is that you are saying, Dee?"

"I wanted to know what you were going to use in place of sphagnum moss. We have the drainage fixed in the box. Shall we put on the coarse material next? Helena said you always used sphagnum moss."

"So I do when I can get it. But I can't always, so I just take some old thin sods and put them on, green side down, next to the drainage. To be sure, the coarse material could go on next, but it is very apt to settle badly in the pot or box. You will find that sod just outside. Pick out the very thinnest pieces; any others will be too thick."

"Now," said Elizabeth, "if you will bring your pots over, we'll fill them up with soil for you."

"Wait a little, Elizabeth. What is going to happen with that pot already full of soil when you put the plant in? Just how are you going to work that?"

"Why, I had intended to make a hole in the soil and put the plant in."

"That is no way to pot plants. Come here, girls, and let us talk this point out. I will pot a plant for you. I guess this begonia would be a good one. See, it has quite a ball of earth of its own. Now look at Elizabeth's full pot. Trying to plant in a pot already full of soil is beginning entirely wrong. Hand over another pot, Josephine. Thank you. See, here is a pot with its drainage, and a very little bit of old sod over this. The soddy matter takes up only about a quarter inch. Give me a trowel full of the potting soil, or a little coarse soil first. Now I lower into the pot my plant with its own earth still about it. See, it is going to be about right. Now, while I hold the plant in position in the pot with my left hand, I shake the potting soil in all about the plant. Here is a stick. I made it and call it my potting stick.

It is about the length of a foot ruler. See, it is about an inch thick and has a rounded end. With this end I gently poke the soil into place. Occasionally, I give the pot a little shake, which settles the soil into crevices and crannies. But never do this jerkily or violently. When the soil is within a half inch of the top of the pot, press it down all about the plant stem; that is, firm the plant. You should be able to take up a potted plant by the plant stalk without uprooting it. The florist can do this with any of his potted plants. If the plant is loose in its new home it will not do well."

"You said to have a half-inch space between top of soil and top of pot. I should like to know why, because all the plants at my aunt's house are done the other way," eagerly inquired Helena.

"The purpose of the half-inch space is for watering the plant. I should be willing to wager that when your aunt waters her plants she has a bad time with water spilling over and soil washing out. The space allows for this and prevents its happening."

"I shall tell her about that when I go to see her. I am to go next week. Don't you think I might carry her a plant nicely potted?"

"Indeed I do. I think we can spare a begonia for her. Just let me water this plant a little. Notice that I do not flood it. Now, set it outside, Elizabeth, right by my back door where it will be sheltered."

"Why put it outside?" asked the child, as she took the pot; "I'd have left it here under shelter."

"It goes out because good fresh air is as important for newly potted plants as good soil, careful handling, and watering. Now for a slipping geranium lesson!

"Old, stocky geraniums in the fall garden are exactly right to slip. These properly slipped and started, if well cared for, will blossom by January or February. If closely crowded into the window box, you may be certain of bloom provided you have good, strong sunlight on them.

Florists slip geraniums and put them into sand; many people put the slips into water to form roots; but it is far easier for you children and for schools to place the slips immediately into the earth of the window box.

"The slipping process itself is easy. Suppose we have a big, old geranium from which to take our slips. It is full of branches. These branches or stems have around them at intervals rings called nodes. The space between two nodes is called an internode. On the nodes are what seem to be small leaves. Press one aside, notice between it and the stem what appears to be a very small bud. Here a new shoot can start.

"So choose a branch, pick off all the large leaves except two at the extreme end. If there are more than two, choose the two smallest leaves. Now it is ready to cut. About four inches down the stem cut it off between two nodes. Do not cut straight across the stem but cut slantwise.

"You have now as I have in my hand, a geranium slip which is four inches long. At one end of the stalk are two leaves; the other end is cut obliquely across. Before you plant this slip look between the two leaves and see if there is a small bud. If so, and it is all green, leave it. For this green means that as the bud develops only leaves will unfold. If you see any other colour in the bud, pinch it out with your fingers. The colour tells us that the bud is a flower bud. If this be left on the slip, all the strength of the little plant will be taken up in forming the blossom. A new plant is not strong enough to stand this. It needs all its power for plant growth.

"Plant the slip in your window box, burying it in earth above the first node; that is, the node just above the cut. Thus you have buried in the earth the place where roots will form.

"Crowd the slips in, three or four inches only apart. They should not be exposed to the full glare of the sun at first. Some gardeners say to let the slips wilt before watering. But it is quite safe to water a little from the first. Do not soak the soil, however, or the young plants will decay.

"Now the slips may be put right into a window box, or into these flats. I would not place them immediately in a sunny window. But place them where the sun does not shine directly. It is excellent to leave window boxes and newly potted plants outdoors for a time, until in the fresh air they get accustomed to their new living conditions."

"Shall we fix up the school window boxes now?" asked Josephine.

"Yes, we will get right at that. But first I will give you a window-box talk.

"The most satisfactory way to grow plants in the schoolroom and in many cases, at home, is in the window box. The window box means the possibility of easily caring for a number of plants in a small space. Plants in pots take much more space than the same number of plants in a window box.

"It is the cleanest way, too. You are all familiar with the sight of a pot covered with crêpe paper stained and discoloured from water spilt upon it and moisture given off from the porous pot.

"The window box, if properly watered, need never leak. Its freshly painted sides need never be covered with any material. It stands for just what it is—a well-made, well-painted wooden box.

"It is quite impossible to give dimensions for the construction of the window box, since it must fit the space one wishes to use. It is wise to keep in mind this—that these boxes when filled with soil are very, very heavy and awkward to handle. So if your window is large, why not have two small boxes for the space rather than one large one? When these are placed end to end the effect is of one long box. The ordinary house window may well have the single box.

"Other things to keep in mind for the constructing of the box are depth, drainage, holes, joints and paint.

"Just as bulb boxes need no great depth, so with window boxes. If the depth be great the plants spend too much energy in root growth. A shallow box means, if properly filled, a compact root mass. So if your box is to be, say three and a half feet long make it not more than ten inches deep.

"As we make drainage holes in the bottom of the bulb box, so we do in the window box. Many people make window boxes without drainage holes. It seems rather better to have them since they offer exit for surplus water, and places for the roots to get at the air. These holes may be bored six inches apart down through the centre of the box; or they may be bored in two lines, thus doubling the number of holes and the amount of air space. Take this rule, for every square foot of space have four drainage holes.

"A box filled with soil all winter constantly in a state of moisture is quite likely to spring or spread apart at the joints. The better fitted the joints the better the box, and the better it stands the inside pressure and moisture continually brought to bear upon it.

"As to paint, of course the box must have one coat (perhaps two) on the outside. A dark green is all around the best. Green is always the best setting for plants. Nature made this colour scheme. We only follow her lead.

"After the box is made, and the paint dry, it must be filled.

"Do any of you girls happen to know just where in the school room the boxes are to be placed?"

"One goes into a south window, and I believe the other is to be in a north one," replied Dee.

"That certainly tells us enough to help us in selecting plants for these boxes. The kind of plants to place in the box depends upon whether the box is to go into a bright sunny window or into one which gets little or no sun. Everyone knows the most satisfactory plant of all for sunny windows is the geranium. It is easy enough to get them for schools without money, because this is the time when everyone's mother is taking up plants for the winter, some of which are always thrown away. Many large, old plants slipped make six or more good little chaps. Begonias are most satisfactory; you can plant these either in sun or shade. A good one for a shady window is the one called the 'beefsteak' begonia.

"Fuchsias make a fine show. If you wish to have a plant of some height do not choose a fern or palm, for these plants need so much water they should always be planted by themselves in single pots or in fern dishes. The amount of water required for a palm would cause ordinary plants, like geraniums, to decay. So choose plants which take up about the same quantity of water. For height, then, one might plant a flowering maple. These are fine in leaf and blossom. So they add much to the box. Dracenas are good for both sun and shade; so, too, is pandanus. The foliage of this is pleasing. Much is added to the general effect if some plants which form long hangers are put in, and planted close to the front side of the box. In sun or shade the Wandering Jew grows. A bit breaks off; it is stuck back into the earth and again it grows. Pieces cut and put into water grow equally well. Trailing over the sides of the vessel they are in, they make a pleasing effect in a corner, or by the side of a window.

"An important thing to be remembered in connection with window boxes is that the closer you plant the better the growth of the plants. It sounds wrong. We think that plants ought to have space to grow in. They should, usually; but space defeats the object of the window box, because the idea is to have top growth and blossom. If you give plants a chance to grow under the ground they will do it at a sacrifice to their growth above ground. So crowd the plants in. The root growth, thus limited and checked, gives added strength above. This is true too, in a measure, of planting in pots. Most people put plants in too large pots, and so fail often to get good top growth and blossom. Notice next time you drop into a florist's shop the large palms in comparatively small pots. Why is this? Just to get good growth of foliage.

"Finally, as to sunny and sunless windows: put into the boxes for the shady windows plants which run to foliage, and into those for the sunny windows plants from which you expect blossom. For blossoms, sun is necessary.

"The last thing of all is the placing of the box. Shall it go on the sill? Not if you wish to keep the sill in good condition. Shall it be screwed to the casement? It may be, but it is hard to place each year, and often the strain is too great on the screws. The best arrangement is that of iron brackets screwed to the casement beneath the window sill. These brackets when not in use may be folded in against the wall and so are quite out of the way and do not have to be removed from schoolrooms each spring when the box goes outdoors. The weight of the box is sufficient to hold the brackets out, and so steadies them that it is not necessary to even screw the box on. Two boys holding the brackets straight, two others placing the box on, is all the labour needed to make that box permanently secure. It remains here now until its journey outdoors next spring.

"I would like to add the English ivy to the list of trailing plants for the window box.

"Some people have candytuft and others marguerites in the indoor box. They do not look as well through an entire indoor season as geraniums, fuchsias and begonias. I think I'll ask Miriam, Elizabeth, and Helena to work on the shady window box. We will use dracena, vincas, pandanus, begonia and Wandering Jew. Ethel, Katharine, and Josephine fix up the sunny window box—the fuchsia, heliotrope, marguerite, geraniums, Wandering Jew, and English ivy. This will be a charming box. Dee, you and I will plant the rest of these geranium slips for the girls."

They all worked away busily for some time. Then The Chief asked the girls to come into the house for a time. As they entered the living room they noticed an array of plants on the big table.

"Sit down, girls, I have a little gift for each of you. I wish you to study and nurse these plants throughout the winter.

"This first tall, rather stiff-looking plant is called an aspidistra. It is the best all-around plant for the house or schoolroom. It does not need much or special care. Be sure to keep the leaves clean. See that you do it, Miriam.

"This little Norfolk Island pine is the only one of the cone-bearing trees that you can bank on. Notice that the method of branching is by whorls. You are to have this plant, Elizabeth.

"I imagined that Eloise would love this little lemon tree called Ponderosa. You can raise lemon trees from seed, but like the apple tree, they need special attention before they grow good fruit.

"The Jersualem cherry tree is for Josephine. It does not absolutely need sun. It, too, stands a great deal of neglect. Remember I am not recommending neglect to you. I am giving you the house plants that are of easiest culture. You will be glad to make note of this entire list. Of course, the berries are the charm of this Jerusalem cherry tree.

"To Ethel I will give this lovely genista. It is the finest of all yellow-flowered winter plants. As the blossoms fade they should be taken off. Since this is a good winter flower, it should be allowed to rest in the summer.

"The azalea is for Dee. It needs the same care as Ethel's plant. These need not be thrown away next spring. But put them in a shady spot laid on their sides.

"Helena's plant is a palm called the *Cocos Weddelliana,* and Katharine's is a fern, the holly fern.

"I wish to tell you a bit about ferns and palms.

"In most school buildings and homes these two kinds of plants hold chief place. This doubtless is because they, too, stand lack of attention. Most people keep them water-logged because supposedly they are accustomed to and need lots of water. We must keep in mind that while ferns for instance are found outdoors in very damp spots, they are not in places undrained and choked off from air. So the jardinière half full of water does not quite represent the real environment of the fern.

"Going on with the fern there are a few points to hold in mind. Do not permit the room temperature to fall below 55 degrees. Neither should it rise above 70 degrees. Direct sunlight injures the delicate fronds of ferns. A north window where there is light without direct sunshine is the right fern place. Keep the leaves clean by spraying with clear water on bright days. If bugs appear, spray weekly with tobacco water. This solution should be very weak.

"The best varieties to grow are the sword, Boston, holly, and maidenhair. Use spider ferns for a fern dish.

"Palms require great quantities of water, even temperatures, little direct sunlight, and daily sponging of the leaves. A sponge dampened in clean water is the best thing to use for this purpose. The most popular and easiest-to-grow palms are the following: *Cocos Weddelliana,* date palm, kentia, and the arcea."

"Will you tell us about the watering of plants?" asked Ethel.

"Surely, although I can say but little since it depends largely upon good judgment. Water a plant when it needs it. Keep the soil moist but not soaked. If plants are beginning to decay or a mould is coming on the surface of the soil, the trouble usually is over watering.

At such a time let the soil of the pot or box dry out. Then water carefully after that. If the pot is always damp or has become green you may know that the trouble is either too much water or poor drainage."

"What pests are likely to attack our plants?" questioned Elizabeth.

"Lice and scale troubles are the chief pests of the house plants. You all know plant lice. Use kerosene emulsion for these. The scale appears as a scaly mass, usually on the stems of plants. These scales are living animal forms. The best way to get rid of them is to wipe them off with a rag soaked in kerosene emulsion."

"Why didn't you give some one a rubber plant?" was the next question from Miriam.

"I guess because of its bad habit of growing so tall and losing its lower leaves. They look like giraffes at the circus. But one may top these plants."

"What is topping?" and "How do you do it?" burst from Dee and Helena at the same time.

"I suppose you see that if one could get that nice-looking top off and start again the old rubber plant would be all right. So about a foot below the last leaf on the stalk—I mean the last leaf numbering from the top—- you should start the operation. Cut a slit in the bark at this place. Pack soil about the stem. Then encase this with sacking. So you have a nice ball of earth packed about the stem. Let the ball be about six inches in diameter. Keep it moist. You can sprinkle the water on. After a time roots will appear coming through the sacking. The roots have started to develop at this incision of the stem. Now the stalk may be cut below these new roots and the new rubber plant potted."

"That is very interesting," began Katharine.

"I should so like to try it," broke in Dee.

"Now, girls, I believe you'd better pick up your new plants and scamper. We certainly have done a good afternoon's work. The chief things to try for in indoor plant culture are cleanliness of the plant, proper drainage, and freedom from abrupt changes in temperature and draughts. Good-by, girls. We meet again soon at our exhibits."

"Good-by, and thank you so much," cried the girls in chorus.

XVIII

THE GRAND WIND-UP—GIRLS VS. BOYS

One day in late September the Boys' Garden Club received an invitation to Katharine's house for four o'clock the following Saturday.

Never were boys more astonished than these as they were ushered into the girls' garden.

"Well," Albert said quite frankly, "you have beaten us as far as beauty goes. You try vegetables next summer, and we fellows will race you."

"Thank you!" answered Dee, "You are very kind to us, very! You may possibly have noticed those beautiful window boxes at the Inn. Eloise made those herself, stocked the boxes, and has cared for them, too. Elizabeth's borders have grown as well, I should say, as Jack's flowers. Peter will tell you of our sundial. Helena has a beautiful garden. To be sure Leston has helped her but only because he wanted to so badly. Girls are not such bad gardeners, are they?"

"Not bad?" shouted the boys, "you are wonders!"

"Every fellow here give three cheers for the girls," commanded Jay. This was done with a hearty good will.

A week later the girls received an invitation to attend the boys' fall exhibit, held at The Chief's house. Early that morning the boys had gone to the woods to gather autumn boughs. The walls were a blaze of bright-coloured leaves. About the room were placed tables upon which the boys' products were exhibited. Fathers and mothers had come to the exhibit; in fact, the whole village had turned out.

The prize went to Peter, for he had made the most money out of his garden.

Just a word from a talk given by The Chief about the arrangement of exhibits and matters in general. He said, "There are as many ways to prepare vegetables and flowers for an exhibit as there are boys and girls to exhibit. It is not enough to bring the articles to be exhibited. The real art of showing one's things comes in artistic and attractive arrangement.

"Vegetables should be thoroughly scrubbed and cleaned. Dirt clinging to the roots needs sometimes a brush to get it entirely off. Carrots, beets, radishes, turnips, celery and other vegetables where the edible part is beneath ground, need this sort of attention, not only to make them clean, but to bring out the colours in each case.

"The foliage of a vegetable plant often adds much to its appearance in exhibition. For instance, the carrot has quite lovely lacey leaves. Beet leaves have such good red colours in them that they, too, add something to the general effect. The colours of the leaf and the colour of the fruit itself are a harmony. When radish leaves are taken off there is no good way of bunching the radishes. They are cut quite off from kith and kin.

"The only objection to the foliage is its wilting, drooping, tired looking leaves certainly add nothing lovely to the exhibit. If the exhibition is of short duration there is no trouble along this line; if it is one of several days the problem is different.

"Children's exhibits, however, usually last but a short time. But if the products can be put away in cold storage over night, or in water in a cool place, then it is possible to keep them in good shape.

"An outdoor exhibit is not wholly satisfactory because of the effect of the air on the products. They wilt badly. A tent is far better than the open-air table for exhibiting. With care the exhibition may always be made attractive.

"The arrangement of products is one of the main features. Generally children's exhibits are pretty messy and mussy looking. This is because of two things: first, the children have so many little separate exhibits; second, we do not stop to discuss carefully the matter of arrangement and preparation. If the children understood fully that no products would be admitted for exhibition unless these were cleaned, were of uniform size, and of the requisite number, there would be little trouble. With them, arrangement would have to be worked out largely on the scene of action, although colour effects, bunching, and general matters could be taken up beforehand.

"Let us think out a few general directions for exhibiting. First, all products must be thoroughly cleaned. Heading vegetables, as cabbage and lettuce, should be cleaned well, and perhaps two heads of each kind shown by the exhibitor. Radishes, carrots, young onions and small vegetables which are to be bunched should have anywhere from twelve to twenty specimens in the bunch. Leave the foliage on such bunches. The large vegetables like beets and parsnips may have from four mammoth specimens to eight smaller ones. Potatoes are exhibited by the plate and so are tomatoes. There are supposed to be seven large specimens to the plate.

"Flowers are usually shown cut and arranged in vases. The vases should be of clear, white glass for the best effects. Rose bowls may be used, too. Do not put grand collections of all varieties and colours of flowers together. Suppose the exhibit of a certain person is to be one of asters. Then put the purple ones together in a vase, the pink ones together in another vase.

"Another mistake of exhibitors is the huddling of products into close quarters. Give your individual specimens plenty of room. Let the things stand out as individual. The entire exhibit is spoiled when it looks messy and huddled up.

"The labelling is often done poorly. Any little piece of paper is stuck on the vases or under a bunch of vegetables. The child's name is written in abominable handwriting. Write or neatly print a little card. Put on this the date, name of the exhibitor (or number) and his place of residence, if required.

"These, in brief, show the real educative lines along which one ought to conduct a children's exhibit. The aesthetic side enters in largely, and a proper bit of the commercial is here, too."

Well, this exhibit of the boys' was pretty good. Each boy had a set of photographs showing the round of his work. These had been made into books. Some of the boys had kept diaries. The diaries had in them not only an account of experiences, but also the tables worked out with The Chief. Jack had what he called an improvement section, which gave ways by which he might improve over his present methods of work. The garden plans drawn to a scale were on the walls. Myron had brought his set of real garden tools. The pieces of hand work made by the boys were there, too.

George had made a collection of garden pests, while Philip and Peter had made collections of weed pests. All the pamphlets from Washington which they had used in their work and those from their own state experiment station were on a little table.

Each boy told briefly the difficulties he had encountered and how he had met them. After a talk by Jay, Albert spoke of the experiment in inoculation of soil. Then he and Jay disappeared, and returned with plates, one for each guest, and on each plate were two spoonfuls of beans, one of the inoculated and the other of the uninoculated beans. The visitors were not told which were which. Then a vote was taken as to which were the better. Of course, the inoculated beans won out.

After this, the real refreshments were served. "I should like to ask," Dee made bold to say, "where you boys got strawberries to make ice cream of? Strawberries in October! You certainly do not expect us to believe you raised them."

"I did," said Myron, striking an attitude before her. "I did in my own little patch."

"Did you make a few cakes of ice and thus have a cold storage plant?" Dee continued sarcastically.

"Dear me, no! I'm much more clever than that. One day, with a few baskets of berries tucked under this noble right arm of mine, I walked to this house. I knocked at the door. A man let me in. He tied an apron about this waist. We actually canned these same berries which you are now eating as a frozen delicacy."

"You boys are altogether too smart," and Dee turned her back on Myron to accept a second dish of cream from Philip.

That didn't disturb Myron any, for he cakewalked back to the kitchen for more cream.

"Well, it was a fine exhibit for mere lads," Jack's father was heard to say on his way home; "If we could bring into this little village a few more men like our boys' Chief there would be no question about a boy's coming up all right. It makes me ashamed to think that we parents have left this work to an outsider."

"I feel," answered George's father, "that this man is a real 'insider'."

After all the guests had left The Chief's once again the boys formed their line and saluted the man.

"Is there nothing for us this winter, O Chief?" asked Albert.

"Plenty. We are going to have a beautiful winter, and next spring better work."

Laden with their trophies the boys reluctantly started for home. They stood in the road in front of The Chief's gate, and the moon shone down on seven happy, manly boys. The three cheers to The Chief arose clear and shrill on the still evening air. As it died away the boys seemed to melt into the shadows of the road.

The man stood motionless in his doorway until the last sound of the boys died away. Then he went back into the room to dream over the fire dreams for his boys.

PART II

THE CHIEF'S GARDEN TALKS

I

THE SOIL

The following winter The Chief gave Friday afternoon talks to his boys and girls. These meetings did not in any way interfere with the boys' regular Saturday evening club.

Immediately after school each Friday afternoon they all trooped round to The Chief's little house, which had become a centre of village interest. Finally the men came too, for they had found out that this man knew of what he spoke.

But we are wandering away from those Friday afternoons.

There was the strangest collection of stools and benches in The Chief's side entry, all belonging to the boys and girls. "You must each one bring your own seat, because you all know that I haven't chairs enough to go around." And this called forth the collection.

It was an odd sight that first Friday in early November. A long straggly line of boys and girls, each one with a seat of some kind, wound its way up to The Chief's hospitable door, where he stood waiting, laughing aloud at the sight. In they came, and made a semi-circle about the big fireplace.

"I just love this room," said Albert, voicing the feelings of them all.

"I have thought," began The Chief, "that since our really successful first year of gardening, we ought to be in a position to undertake and to desire to know more about certain subjects which I shall discuss. Each Friday I am going to take up a topic such as I should if I were teaching you in school."

"You do not mean that we'll have to remember and answer questions just like school? You surely do not mean that, Big Chief," broke in Albert.

"No," replied the man laughing, "no, you may forget it all if you like. Remember it, if it seems to you useful. But if it's a strain on you, Albert, make it your business to forget."

They all laughed at this, but none so heartily as Albert himself. "That's one on this old head of mine," he said, banging that member up against the side of the chimney.

"My first talk I have given you in part, but I have more I wish to add. I believe even Albert can stand it. The subject is the soil.

"Soil primarily had its beginning from rock together with animal and vegetable decay, if you can imagine long stretches or periods of time when great rock masses were crumbling and breaking up. Heat, water action, and friction were largely responsible for this. By friction here is meant the rubbing and grinding of rock mass against rock mass. Think of the huge rocks, a perfect chaos of them, bumping, scraping, settling against one another. What would be the result? Well, I am sure you all could work that out. This is what happened: bits of rock were worn off, a great deal of heat was produced, pieces of rock were pressed together to form new rock masses, some portions becoming dissolved in water. Why, I myself, almost feel the stress and strain of it all. Can you?

"Then, too, there were great changes in temperature. First everything was heated to a high temperature, then gradually became cool. Just think of the cracking, the crumbling, the upheavals, that such changes must have caused! You know some of the effects in winter of sudden freezes and thaws. But the little examples of bursting water pipes and broken pitchers are as nothing to what was happening in the world during those days. The water and the gases in the atmosphere helped along this crumbling work.

"From all this action of rubbing, which action we call mechanical, it is easy enough to understand how sand was formed. This represents one of the great divisions of soil—sandy soil. The sea shores are great masses of pure sand. If soil were nothing but broken rock masses then indeed it would be very poor and unproductive. But the early forms of animal and vegetable life decaying became a part of the rock mass and a better soil resulted. So the soils we speak of as sandy soils have mixed with the sand other matter, sometimes clay, sometimes vegetable matter or humus, and often animal waste.

Constant Cultivation of the Soil Saved George's Cabbages

Photograph by Karl W. Helmer

"Clay brings us right to another class of soils—clayey soils. It happens that certain portions of rock masses became dissolved when water trickled over them and heat was plenty and abundant. This dissolution took place largely because there is in the air a certain gas called carbon dioxide or carbonic acid gas. This gas attacks and changes certain substances in rocks. Sometimes you see great rocks with portions sticking up looking as if they had been eaten away. Carbonic acid did this. It changed this eaten part into something else which we call clay. A change like this is not mechanical but chemical. The difference in the two kinds of change is just this: in the one case of sand, where a mechanical change went on, you still have just what you started with, save that the size of the mass is smaller. You started with a big rock, and ended with little particles of sand. But you had no different kind of rock in the end. Mechanical action might be illustrated with a piece of lump sugar. Let the sugar represent a big mass of rock. Break up the sugar, and even the smallest bit is sugar. It is just so with the rock mass; but in the case of a chemical change you start with one thing and end with another. You started with a big mass of rock which had in it a portion that became changed by the acid acting on it. It ended in being an entirely different thing which we call clay. So in the case of chemical change a certain something is started with and in the end we have an entirely different thing. The clay soils are often called mud soils because of the amount of water used in their formation. The slate that Myron brought for road making belongs to the clay family, and so does shale.

"The third sort of soil which we farm people have to deal with is lime soil. Remember we are thinking of soils from the farm point of view. This soil of course ordinarily was formed from limestone. Just as soon as one thing is mentioned about which we know nothing, another comes up of which we are just as ignorant. And so a whole chain of questions follows. Now you are probably saying within yourselves, how was limestone first formed?

"At one time ages ago the lower animal and plant forms picked from the water particles of lime. With the lime they formed skeletons or houses about themselves as protection from larger animals. Coral is representative of this class of skeleton-forming animal.

"As the animal died the skeleton remained. Great masses of this living matter pressed all together, after ages, formed limestone. Some limestones are still in such shape that the shelly formation is still visible. Marble, another limestone, is somewhat crystalline in character. Another well-known limestone is chalk. Perhaps you'd like to know a way of always being able to tell limestone. I'll drop a little of this acid on some lime. See how it bubbles and fizzles. Now Albert will drop some on this chalk and on the marble, too. The same bubbling takes place. So lime must be in these three structures. One does not have to buy a special acid for this work, for even the household acids like vinegar will cause the same result. Albert will prove this to you.

"Then these are the three types of soil with which the farmer has to deal, and which we wish to understand. For one may learn to know his garden soil by studying it, just as one learns a lesson by study.

"I believe the boys from their last winter's work feel fairly familiar with soils, I have in these three tumblers the three types of soil. As I pour water on them just see what happens. Observe how little water it takes to saturate sand. The limy soil holds more water and the clayey an amazing quantity.

"I do not know whether you are much acquainted with the sea shore, I doubt it."

"I am," broke in Katharine, "for each summer, except this last one, I have spent a month at the beach."

"Then possibly you can tell us, Katharine, whether, or not, the sand takes in, or absorbs, much heat during the day."

"Indeed it does absorb heat; why some days we used to go barefooted on the beach right after dinner. I can tell you there were times when we couldn't stand the heat of the sand."

"That is quite true," continued The Chief, "sand absorbs heat to a remarkable degree. This heat is, to be sure, in the upper layers of the sand. Had Katharine burrowed down with her toes below those upper layers she would have found moist, cool sands. But an upper layer of soil, made up of particles which fall apart easily because of the loose make-up, a layer which has absorbed little water and much heat—well, to me that sort of soil doesn't sound quite right for good gardening. Add to such a soil, humus in the shape of stable manure in large quantities and this same poor soil becomes very good.

"Now here is the lime soil tumbler. This soil has taken up rather more water than the sand took. But it, too, surely needs to develop greater power to take in and hold water. So the same sort of medicine which we gave the sandy soil may be dealt out to the lime soil. Lime is a pretty good substance to have in soil. Lime is a kind of fertilizer in itself; it's a soil sweetener; it helps to put plant food in shape for use, and causes desirable bacteria to grow. This sounds a bit staggering but all of these things I am going to talk over with you. So just at present forget it, Albert, if it is a heavy burden.

"The clay soil, you observe, has taken in quite a quantity of water. That seems like a good thing. It is. But clay has a mean little habit of squeezing tightly its particles together with the aid of water so that air is excluded from the mass. It forms huge lumps; it bakes out and cracks badly; and it is also very damp, cold and soggy in early spring.

"As the problem with sand is to add something so that more water may be held in the soil, so the problem with clay is to overcome that bothersome habit of baking and caking and cracking. To do this we might add sand or ashes. But perhaps it would be better yet to add manure with a lot of straw in it. This is the easiest kind of thing for country boys and girls to get, because the bedding swept out of horses' stalls is just the thing.

"When I speak of clay's horrid habit of tight squeezing, I always have to stop and talk about the two great needs of all soils. One is the need for water; the other, for air. A soil cannot exist without these two things any more than we can. Without these, or poorly supplied with them, a soil is as if it were half-starved.

"That trouble always comes from a lack of one or the other is quite sufficient to prove to us that these are essential. Just see how sand lacks water, as does lime soil too! But there is plenty of air space, unless these soils are too finely powdered. Now look at clay! plenty of water, but how about the air? When clay begins its packing, then air is excluded.

"So one of the questions to be asked in soil improvement concerns the water and air problem. We must have air spaces, and we must have water-holding capacity.

"Before we go home I must just speak of soil and subsoil. When you strike your spade down into the earth and lay bare a section of the soil this is what you see: on top is the plant growth, the soil beneath this, dark in texture and about our locality of a depth of from six to eight inches. This layer is called the topsoil. In sections of the West it is several feet in depth. Now below the topsoil is a lighter coloured, less fertile, more rocky layer, the subsoil. Beneath comes a layer of rock.

"And finally you may be a bit confused by the word loam. It is often given as one of the classes of soils. By loam we mean clay, sand and humus. You will often hear people speaking of a sandy loam or a clayey loam according as there is a greater percentage of sand or clay in the soil.

"Next Friday I shall talk about soil fertility. So trot home lively now!"

II

PLANT FOOD

A soil, as I have said before to the boys, may contain all the food necessary for plant growth and still not support any good growth at all. That means then we ought to be able in some way or other so to understand the soil that it will be possible to unlock these good things for the plants to live on.

"I see a question in Josephine's and Miriam's faces. I guess that this question is concerning what the plant food is in soils. That is right, is it not?

"Well, I'll take that up first, then;—different ways of improving and increasing the goodness of the soil.

"The foods that are necessary and essential to plants and most likely to be lacking in the soil are nitrogen, potash and phosphorus. Now by no means must you think that these are the only chemicals which are foods, for there are something like thirteen, all of which do a share in the food supply. Oxygen and carbon are very necessary indeed. Oxygen is both in the air and in water. Carbon plants take entirely from the air. I might go on and tell you of iron, of sulphur, of silicon and all the others. But you would only get confused, so I am going to make you acquainted with these three entirely necessary ones. They are capricious; often missing, and when not missing hard to make into available food for plants.

"The soil contains many bacteria, small living organisms. These may be divided into two classes, the good ones and the bad ones. The good ones acting on nitrogenous matter put it in shape for the plant to absorb or feed upon. You see nitrogen may be in soil in quantities sufficient for nourishment. But unless it is in a compound available for use, it is of no value to the plant. Then there are the bad bacteria which act upon nitrogen in such a way as to form compounds which escape from the soil as a gas. That is pretty bad, is it not?

"How can the good bacteria be encouraged to grow, and the bad ones prevented from forming? The necessary conditions for the growth of good bacteria are air, water, darkness, humus matter and freedom from acid condition of the soil. If the soil is acid then these other 'chaps' set up their work; so we must see to it that our soils are well cultivated, well aired, have plenty of manure, and, if acid, have a liming, so that these bacteria missionaries can start their good work.

"The manure I spoke of above is the great source of nitrogen upon which most plants depend. There is nitrogen sufficient right in the air, but that again is not available. Certain plants like beans, peas and clovers belonging to the family of legumes are a great deal more fortunate than the rest of the plant families, for, under favourable conditions, they develop bacteria which make it possible to take into themselves free nitrogen. Just look here! See this narrow box; I can drop down one side of it. Here is a sheet of glass put on so you may look at the roots of the beans which are planted close to this glass side. Just observe the great extent of root system. Now see on the roots these white lumps, or nodules as they are called. These contain nitrogen-gathering bacteria. Some farmers in order to get more available nitrogen in the soil plant a crop of some legume. Then these root masses with their treasures on them are spaded into the soil.

"But most plants depend for nitrogen on manure. Whenever you see sickly looking foliage know that nitrogen is lacking, and supply manure in order to obtain it.

"The next element is potash. Its most common source of supply is wood ashes, not coal ashes. One may buy potash in the form of the muriate or sulphate. I told the boys before that potash was good for seed and fruit. Pretty necessary to have in the soil, is it not? Stunted fruit and poor seed mean lack of potash. Phosphorus helps in this work too, and also assists in the forming of fine flowers. Bone ash and phosphates are the sources of this food element.

"So if we just consider the classes of soils with which we have to deal, remember the foods that must be had, and the effects on plants where one (or all) of these is lacking, we have in our hands a help to soil troubles.

"Take sandy soil—what is its greatest need? I should say humus. It certainly should have more nitrogen. So add humus in the form of manure. Spread it on your piece of garden plot anywhere from two to six inches deep. This spaded in will, I think, do the work. You see sand allows water to trickle away too fast. Water must be held properly in the soil.

"The clay soil really needs air. The good bacteria will not work without this. So spade the soil up in the fall, and leave it weathering in huge lumps. Sand or ashes added in the spring helps the air question too. A sprinkling of lime over the surface tends to sweeten the entire soil; for clay soil, so often too wet, is liable to get sour. Lime also adds another plant food called calcium. It would not be bad to add some humus in order to have an even greater supply of nitrogen.

"The lime soil, light and sweet, needs humus too. It should have this to add body and ability to hold water.

"Sometimes it is well to add in the spring a sprinkling of phosphates; that is a chemical fertilizer. Chemical fertilizers are like tonics to the soil.

"All this very briefly puts us in touch with plant foods. I think you all know from your school work that plants take their foods in liquid form. These solutions of foods are very, very weak. That is another reason why we should see that, if possible, there is plenty of nourishment available in the soil, and plenty of water too.

"These bean roots and rootlets show the feeding area or places of plants. Notice the small roots which apparently have a fringe on them. These fringes we call the root hairs. These absorb, soak up the dilute food which is in the soil.

"It is very wonderful what power they have of penetrating the soil. See the bit of blotter I have put down the path of one bean's root course. It would seem to shut the rootlets entirely off from the soil.

"Jay will gently press the bit of blotter away from the soil. See here and there how these root hairs have wound their way through the blotter to the soil, their feeding place. It is well that plants have this power of seeking and finding food. Because it greatly increases their food chances.

"So much very briefly for plant food. I have not told you very much to be sure, but it is quite enough, I think, for a 'starter,' I wish to tell you a bit about the plant itself soon. A few experiments may liven up the subject. So I shall ask Josephine, Miriam, and Ethel to attend to those for us. We can take turns at demonstrating as Jay and Albert have to-day. So you girls must remember to drop in to see me—say, Wednesday of next week."

III

SEEDS

Now before we begin just have a look at these geraniums. They have turned entirely around again and are looking out of the window at the sun. The power which plants have to move is very clearly shown, is it not? I am going to talk a little this afternoon about seeds.

"Any reliable seed house can be depended upon for good seeds; but even so, there is a great risk in seeds. A seed may to all appearances be all right and yet not have within it vitality enough, or power, to produce a hardy plant.

"If you save seed from your own plants you are able to choose carefully. Suppose you are saving seed of aster plants. What blossoms shall you decide upon? Now it is not the blossom only which you must consider, but the entire plant. Why? Because a weak, straggly plant may produce one fine blossom. Looking at that one blossom so really beautiful you think of the numberless equally lovely plants you are going to have from the seeds. But just as likely as not the seeds will produce plants like the parent plant.

"So in seed selection the entire plant is to be considered. Is it sturdy, strong, well shaped and symmetrical; does it have a goodly number of fine blossoms? These are questions to ask in seed selection.

"If you boys and girls should happen to have the opportunity to visit a seedsman's garden, you will see here and there a blossom with a string tied around it. These are blossoms chosen for seed. If you look at the whole plant with care you will be able to see the points which the gardener held in mind when he did his work of selection.

"Last winter we had quite a discussion on corn seed selection. So we will not discuss that further. Only let me say this for the benefit of the girls in order to show them the care which must be exercised in selection. Should a finely formed ear of corn have one or two black kernels on it, then that shows a cross or taint, do not use such an ear for the old trouble may crop out. Take an ear of seed corn, notice the small and rather undersized kernels at the top; do not use these. Select kernels, the largest, plumpest and best shaped.

"In seed selection size is another point to hold in mind. Suppose Peter had bought a package of bean seed. Pull the little envelope out of your pocket, young man, and open it up. Just look at those seeds as Peter spreads them out here. Now we know no way of telling anything about the plants from which this special collection of seeds came. So we must give our entire thought to the seeds themselves. It is quite evident that there is some choice; some are much larger than the others; some far plumper, too. By all means choose the largest and fullest seed. The reason is this: When you break open a bean—and this is very evident, too, in the peanut—you see what appears to be a little plant. So it is. Under just the right conditions for development this 'little chap' grows into the bean plant you know so well.

"This little plant must depend for its early growth on the nourishment stored up in the two halves of the bean seed. For this purpose the food is stored. Beans are not full of food and goodness for you and me to eat, but for the little baby bean plant to feed upon. And so if we choose a large seed, we have chosen a greater amount of food for the plantlet. This little plantlet feeds upon this stored food until its roots are prepared to do their work. So if the seed is small and thin, the first food supply insufficient, there is a possibility of losing the little plant.

"You may care to know the name of this pantry of food. It is called a cotyledon if there is but one portion, cotyledons if two. Thus we are aided in the classification of plants. A few plants that bear cones like the pines have several cotyledons. But most plants have either one or two cotyledons.

"Some plantlets, as they develop and start to push above the ground, bear along the cotyledon. This is true in the case of the bean. Jack and Peter have planted corn and beans in this box, not to have succotash but to show you about the habits of seeds. See the bean plantlet, big, sturdy, fellow, is still clinging to its seed leaves or cotyledons, its baby nourishment. Now look at the corn: there is absolutely no sign above ground of its one seed leaf.

"So from large seeds come the strongest plantlets. That is the reason why it is better and safer to choose the large seed. It is the same case exactly as that of weak children. Look at Myron, great strapping fellow! Hasn't he a fine chance in the world? Do you remember that little sickly boy who was in school last spring? He was as old as Myron, yet see how handicapped he is. Try not to bring weak little plantlets into the plant world. Bring strong, sturdy, healthy ones by careful seed selection.

"There is often another trouble in seeds that we buy. The trouble is impurity. Seeds are sometimes mixed with other seeds so like them in appearance that it is impossible to detect the fraud. Pretty poor business, is it not? The seeds may be unclean. Bits of foreign matter in with large seed are very easy to discover. One can merely pick the seed over and make it clean. By clean is meant freedom from foreign matter. But if small seed are unclean, it is very difficult, well nigh impossible, to make them clean.

"The third thing to look out for in seed is viability. We know from our testings that seeds which look to the eye to be all right may not develop at all. There are reasons. Seeds may have been picked before they were ripe or mature; they may have been frozen; and they may be too old. Seeds retain their viability or germ developing power, a given number of years and are then useless. There is a viability limit in years which differs for different seeds. This matter, along with directions for testing, the girls may get from our club secretary, Albert. All of this we took up last year in our preliminary garden work before we started outdoor work.

"From the test of seeds we find out the germination percentage of seeds. Now if this percentage is low, don't waste time planting such seed unless it be small seed. Immediately you question that statement. Why does the size of the seed make a difference? This is the reason. When small seed is planted it is usually sown in drills. Most amateurs sprinkle the seed in very thickly. So a great quantity of seed is planted. And enough seed germinates and comes up from such close planting. So quantity makes up for quality.

"But take the case of large seed, like corn for example. Corn is planted just so far apart and a few seeds in a place. With such a method of planting the matter of per cent, of germination is most important indeed.

"Small seeds that germinate at fifty per cent. may be used but this is too low a per cent. for the large seed. Suppose we test beans. The percentage is seventy. That per cent. would pass you in school, but it does not pass muster here. For if such low-vitality seeds were planted, we could not be absolutely certain of the seventy per cent. coming up. But if the seeds are lettuce go ahead with the planting. Peter will pass around these germinating per cent. tables which he has printed for you. I'd advise you to paste these in your garden diaries. After a test refer to this table which is from a United States Agricultural Dept. list for seeds not over one year old. You then know at once whether the seed is worth using.

PER CENT. OF GERMINATION

Beans	90
Cabbage	90
Carrots	80
Cauliflower	80
Celery	60
Corn	87
Cucumber	87
Eggplant	80
Leek	80
Lettuce	85
Muskmelon	87
Okra	80
Onion	80
Parsley	70
Peas	93
Peppers	80
Pumpkin	87
Radish	90
Spinach	84

Squash	87
Tomatoes	85
Turnips	90
Watermelon	87

"After being sure of good seed the next step to consider is when to plant the seed. It is well to start certain seed inside and so get a bit ahead of the season. Other seed may as well wait, and be planted out in the open when the ground is warm.

"Such vegetable seed as the following may be started inside.

INDOOR PLANTING TIME-TABLE

FEBRUARY	MARCH	APRIL
Artichoke		
Cabbage		
Celery	Cucumber	
Egg plant	Egg plant	
Lettuce	Lettuce	Muskmelon
Onion	Pepper	
Radish	Radish	
Tomato		Tomato

"Flower seeds I will take up later because I wish to think over the flower garden by itself.

"When shall we plant seeds outdoors? Now no one under the sun can say plant such and such a seed on May 30th or April 1st. It is the same absurd case as saying change your winter clothes for summer ones on May 1st. Many writers will cover this subject by saying plant seeds when the earth is warm. But even that is a pretty general sort of direction.

"Nature has given us a planting guide. She tacks her notice on the fruit trees. When those early blooming trees, the peach and the plum, put out their beautiful blossoms the first planting time is on. To be sure the temperature then is a bit low, only about 45 degrees, so the planting is not of the more tender vegetables. Get your seed of beet, carrot, cabbage, cauliflower, endive, kale, lettuce, parsley, parsnip, onion, pea, radish, turnip and spinach. These may all be planted.

"The next signal to watch for is given by the blooming of the apple trees. This is the planting time for the more tender seed. These need a temperature of about 60 degrees in the shade, real apple-blooming time. Corn, beans, egg plant, melon, squash, cucumber, pumpkin, tomato and pepper seeds may be planted.

"But when is the time to put out the hotbed, or indoor-started seedlings? When the apple blossoms drop their petals and have passed by is the signal for them to go into the ground. Of course, they naturally would be the last, for they are made very tender from their glass-grown coddling.

"When it comes to the planting of seed there are certain things to remember always. First the ground should be made very fine. This is an easy matter if the planting is done in the hotbed, but more of a problem in the outdoor garden. It is foolish to plant at all if one does not intend to do things right. So work over the seed bed thoroughly. After all is fine and deeply worked, say to about a foot deep, the next thing to consider is this—how deep should a seed be planted?

"The depth depends upon the size of the seed. Take such small seed as poppy, parsley, even lettuce, and these may be just sprinkled on the surface of the ground. Then tread them in with the foot or place a board over them and walk on the board. In this way the small seed are pressed into the soil quite sufficiently.

"For seeds in general the following might answer for a rule: There are seeds like corn, oats, wheat and the grasses which come up unhampered by their seed leaves. Such seed may be planted deeply—say ten times the thickness of the seed. Other seed like beans, squash, radish, etc., push and carry their seed leaves up through the soil with them. So these, because of this extra work, should be planted nearer the surface. Four or five times the thickness of the seed is a safe rule to follow.

"When the seed becomes entirely or nearly saturated with water then germination begins. Sometimes people soak their corn in tepid water before planting. This hastens germination. But on the other hand if the soil is very wet and cold the soaked seeds may rot in so much moisture. Certain seeds have very thick coverings. Canna, date and nut seeds are examples. Their cases are so hard and absorb moisture so slowly that germination is a long process. To hasten this little holes may be drilled in the case, thus giving the plant germ a chance to get out. Nurserymen crack the nuts in order to help matters along. You can readily see what a really difficult piece of work it is for a tiny embryo or baby plant to break open a thick case.

"If seeds are planted too deeply again, a tremendous piece of work is imposed upon the little plant. To push up through, say one inch of soil, would be quite a task for a lettuce seedling.

"Finally in seed planting, the soil must be safely compacted or pressed about the seed. The object of this is to bring in contact with all parts of the seed soil particles with their films of water. Suppose a radish seed is planted and no soil happens to come in direct contact with the seed. That distance, so slight to us, is a well-nigh impossible one for the rootlets to extend to.

"There is a possibility of course, of too close compacting. This occurs when the soil is very wet. Do not compact at all then. In fact, such soil condition represents a very bad time for planting, anyway. Moisture is necessary for germination, but superabundance of water is fatal. It is simply ideal when after a planting a gentle rain comes—germination.

"I remember once seeing a garden which school children had planted so close to the surface that after a rain most of the seeds were lying all sprouted on the surface of the soil. Take care not to plant in such a manner.

"This talk has been largely for the purpose of bringing to your minds certain necessary points. Let me sum them up: Cheap seed are expensive because they are often full of impurities and lack vital power. Buy good seed and test *them*. Plant large seed, because the storage of food is greater. Make the soil conditions right in order to give every help to the seed. Plant neither too deep, nor too near the surface. Compact the soil, and so aid germination. The first start of work must be right; otherwise, trouble comes."

IV

THE PLANT ITSELF

"To think of a plant as a breathing, growing thing is wonderful, but it is far more wonderful to think of it as something possible for even boys and girls to train and improve. Here is a bed of petunias, let us say; do you know just how it is possible to have larger, finer petunias next year?

"A slight operation performed, and behold magic has been worked!

"First, we will go over the life history of a plant, and then I'll tell you of this magic and how to work it. Or better yet my assistants here, Josephine, Miriam and Ethel, will do the trick.

"A plant really goes through much the same operations in life as does an animal. Only to be sure, these operations are performed in a rather different way. A plant has a digestive, or feeding, system, a breathing apparatus, the power to rid itself of waste and to make seed; it moves, and it grows, too. Philip looked a bit skeptical when I said it moves. Well, it does. Of course, a plant does not walk about, and move from spot to spot. But a plant can and does move. Why it can turn itself around back to, even. Just look at my geranium slips there! they seem to be breaking their backs to peep out of the window and look at their best friend, the sun. Turn all of them around, George. See, they face us now! remember to look at them next Friday.

"But to start over again. A plant has just three necessary and important parts: these parts are the roots, stem, and leaves. No, Elizabeth, the fruit and flowers are not separate parts. Why? Well, merely because by some queer provision of the plant world, the leaves are responsible for making or forming both the flower and the fruit. If you watch a bud form and unfold, you will notice that the entire little bud seems to be a series of leaves. And if your fingers were clever enough you could take tiny leaves and fold them into the parts which go to make up the flower and the fruit. This last, like most of the rest of that I am telling you, is just one of the miracles of nature.

"The root, rootlets and root hairs all go to make up the root-system of a plant. This system is a feeding and food storage system; cold storage, we might call it.

"I have spoken before about how the root hairs absorb food. Food is soaked up something as a blotter soaks up ink. Underground plant food must be liquid in nature. This is because plants, like babies, must have very dilute food. Plants can no more get food out of a dry lump of soil than a little baby can get its food from a hunk of bread or a thick slice of corn beef. But let that soil be water-soaked, and have the proper bacteria at work, and the material is in plant-food form. Josephine has here an old, old experiment. What was a white pink is now a red one. It has been in that glass of red ink and a little water. And lo, up the stem the red fluid climbed until it suffused the white flower and made it red. Notice as Miriam holds that lump of sugar only just touching the surface of the water, the water moves up that lump. In this way water and liquid food rise up the stems of plants. Just so, too, water rises in the soil from the lower layers up to the feeding place of the roots, and even up to the surface of the ground.

"As the roots are feeding and storing places, so the stem is a sort of passage way for the passing back and forth of liquids. Take a stem of a big plant, like an oak tree, and you see in the wood where storage of fibre has gone on. But the great work is that of interchange.

"Leaves are very active portions of the plant. They represent a great, busy manufactory. Manufacturing what? That question I see stamped on Myron's face so plainly he need not speak it out. Manufacturing real food out of raw material—that is the work of these plant shops.

"Let me tell you about this. Ethel has in her hands two little plants. The one in her right hand has been growing in the light; the other, in her left hand, has been put away in the dark to grow. The absence of green colour is very marked in this latter plant. So you see it takes light to form this green, or chlorophyll as it is called. The chlorophyll-saturated cells, absorbing carbonic acid and the water-diluted food from the soil, literally break them up. And when broken, food is found suitable for plants to absorb. Wonderful, is it not?

"I spoke of carbonic acid; well, this is a gas, as some of you have found out before, made up of carbon and oxygen. It is a gas which we of the animal kingdom breathe out as waste from our bodies. The plant takes it in through the leaf—and, by the way, I ought to explain that. It is this way: if we had a magnifying glass we should find over the inner surface of leaves, pores, or stomata as they are called. They open in the presence of light; and from these openings what the plant has no use for passes out, and gases from the air may pass in. Some call these openings breathing pores.

"Quantities of water pass out through these pores. When this process goes on too rapidly a plant will wilt.

"So, to go back, we will suppose that carbonic acid gas has passed into the leaves. Straightway the chlorophyll bodies get to work. The gas is broken up, and oxygen and carbon are left. The carbon is wood the plant builds. Some of the oxygen passes out into the air and some is kept for plant food use.

"It is a good thing for us that some of the oxygen does escape into the air for we need it. So you see we, in our respiration, and the plant, in its breathing, are doing each other a good turn.

"Of course, there is the dilute food from the soil, which is largely mineral matter and water. The chlorophyll bodies work away on these minerals, and make them into foods. A great body of water, as I have said before, passes out of the plant through the stomata.

"I have told you a thing that the plant can do which we are not capable of doing. A plant takes a mineral and makes it over into food. You and I, unless we happen to be circus glass-eaters, are not built to do this work. But the vegetables which we eat do the work for us.

"A great deal of plant food is in the form of sugars and starches. I remember Katharine and Peter told me last winter that in their physiology they learned how sugars and starches were made in our own bodies. And lo and behold, the geranium can do a similar thing.

"Some plants store up lots of starch, as the potato. Others store quantities of sugar, as the Southern sugar cane and the beet. Wonderful? Well, I guess it is. If we could hear and see all the work these energetic little chlorophyll bodies are doing, we should be amazed.

"You will remember that I told you some plants could take the very necessary chemical nitrogen from the air; most of them, however, must get it from the soil. And so again this from the soil solution is worked over into available food.

"After all we must not fail to see that water is most important. It floats all the important food elements to the leaves for the work to be done there. The food carbon, of course, is an exception to this rule and I will say again in certain cases nitrogen is, also.

"Thus you boys and girls now understand how necessary it is that a soil should be of the right texture to hold water. If it is not, it must be helped to be so. Sand, you will remember, had to be doctored to hold water. Clay needed treatment in order to make it quit its bad habit of baking out.

"Here is a rather interesting experiment set up by Josephine and Ethel. Look at the first piece of apparatus—a tumbler partly full of water, a piece of cardboard over the top of the tumbler, and passing down through a hole in the cardboard a piece of plant just stem stripped of leaves, and finally a second tumbler clapped over the first. The second piece of apparatus is exactly like the first, only that the stem, one end of which is in the water, has leaves on the other end. Notice that the upper glass in the second case has moisture on it. The upper tumbler of the other set is perfectly dry. Whence, then, came the moisture? It must, of course, be the leaves which gave it off, since they represent the only difference in the two pieces of apparatus.

"I wish we might go on with whole sets of experiments, but for that we have not time.

"You understand a little of the mission of root, stem and leaf. The root does a good work in holding a plant in place. It is the foundation material of the plant. There is much, much more to be learned about all these subjects. This little is just to open your eyes to the wonders of the work each plant is performing all the time.

"I said I would show you some magic. Well, this magic has to do with plant improvement. It is not much of a trick to raise a plant, but it is a great one to be able to improve that plant.

"Let me tell you of a friend of mine whom we will call Rodney, because that is his real name. One day Rodney noticed the gardener doing something with a little flat knife to a pansy. Then he tied a little paper bag over the pansy, of course leaving the whole thing on the plant.

"'What are you doing?' asked the lad. 'I am fixing that pansy so that the seed from it shall be finer seed than they otherwise would be.'

"Then the old gardener explained this to Rodney: There are two parts to flowers which are very necessary, absolutely necessary to making seed. One part is the pistil, the other the stamen. Some flowers have both pistils and stamen, while others have just the pistil and one has to hunt for another plant having the stamen. You can tell the stamens in this way: they are the parts which have in their care the pollen. Most of you know pollen as a yellow powder or dust. Sometimes it is a sticky gummy mass. The pistil is that part of the flower which ends in the seed vessel. It very often takes a central position in the flower, standing up importantly as if it were the 'part' of the flower. And after all, it is. Now, when this pollen powder falls on the pistil it does not explode. The pistil merely opens up a bit and down travels the powder into the seed vessel to help form seed. There would be no real fertile seed without the pollen.

"Sometimes the pollen from one flower falls on its own pistil, sometimes the wind, the bees, the birds carry the pollen to flowers far off and drop it on their pistils. Marvelous, is it not? Everything has to be just right, or the pollen does not do its work nor the pistil, either. Pollen has to be ripe to help make the seed.

"But how can the work of the wind and the bees and the birds be improved on? Just as the old gardener was doing it. He had one pansy, oh such a large one, but not at all beautiful in colour. He had another one, small but exquisite in colouring. If he could but grow those two together, shake them up, say a magic word and get a pansy both beautiful and large!

"Rodney's gardener used magic but not a magic wand. He took a little knife called a scalpel. He carefully took some pollen from the beautiful pansy and then rubbed it gently over the pistil of the big pansy. The pollen was all ready to drop, and by this he knew it was ripe.

"Why did he place a bag over the pansy? Well, simply because he didn't wish that pansy interfered with. Suppose the bag were not on; suppose after he had put the pollen on, the wind had blown other pollen to this same pistil? Let us suppose that this other pollen came from a very inferior flower. The experiment would have been spoiled.

"Any of you can try this plant improvement. I see by Katharine's eyes and Dee's also that they are going to try it. It is well if you have a pair of forceps. Then you need not use your fingers against the plant at all. Gently pull the pistil a bit forward, gently place the pollen on with the scalpel and you have performed the operation entirely with the proper instruments.

"The girls did some saving of fine specimens of flowers this fall, but the kind of work of which I have just told you means far more. In the one case you choose from what you have; in the other case you make what you want.

"Good-by, again, until next Friday afternoon!"

V

INCREASING PLANTS

"This last garden season we have tried two methods of raising plants: one was by seed; the other by slips or cuttings. The girls will typify still another method with their bulbs. This last method is by division. A bulb as it stores up its nourishment after the blossoming time forms new little bulbs. These may be separated from the parent tuber if large enough. You all saw me dividing my peonies. Those peonies doubtless were started years ago from one or two roots.

And now when I dug them up it looked as if I were laying in a stock of sweet potatoes so great was the increase.

"There are just three other methods of propagating or increasing plants. These are layering, budding and grafting.

"Layering is done in several ways. Suppose you have a gooseberry bush you wish to layer. The time to do the work is after the flowering season is past. Choose a branch which has not flowered. Strip off the lower leaves. Now where the old and new wood meet is the place for the cut. Make a cut right into the stem which will be like a tongue. Let this be about an inch long. Hold this to the ground with the cut side down. Bank soil over this. At and under the tongue the new shoots will start, and the new gooseberry bush grow from this. This new plant may be cut off from the parent. If the twig will not stay bent down in this position, cut a forked piece of wood which shall act as a pin. Do you picture this? A branch bent so that not far from the parent plant it is buried under ground with the rest of the root protruding from the ground.

"A rubber plant may be layered or topped as it is called. Rubber plants have an ugly habit of going to top, dropping off their lower leaves as they do this. So they look as if they were trying to develop into huge bushes, and they become very ugly in so doing. The top looks all right and many a person wishes that top were off all by itself and nicely potted.

"This is the way it is topped. A slit is cut in the bark about where you would like to see roots growing. Then soil and florists' moss is bound about the wound. These may easily be kept moist. A paper pot could be put about the soil if one wished. The soil mass should be a ball of about six inches in diameter. When the new roots appear through the moss or poking out of the paper pot, cut the stem of the plant below the pot. And behold you have a little rubber plant just as good as new, I have told this before to the girls.

"Another method of layering is to cut the parent off down to the ground. What is left is called the stool. This stool should be covered with about six inches of earth. Let us suppose this is done in early spring. When fall comes around uncover the stool. There will be found a number of new shoots or plants all nicely protected. These may be transplanted. Do you know that stool can be used over again?

"This work of layering is really very simple. Myron used it with his strawberries. The runners were bent and buried just as those of the gooseberry I spoke of. In this way new strawberry plants were obtained. One shoot may be bent and buried more than once. So one may get just as easily two or more new plants from one shoot. This seems as much a miracle as the cross fertilization of plants.

"The fifth method is that of budding. Apple, peach, plum, cherry and pear trees may be budded; so, too, may roses.

"In a word, a bud is taken from some desirable tree and inserted within the bark of a tree either less desirable or young. Young fruit trees, as you know, need some help before they produce good fruit. Now if George had at home a peach tree which bore very fine fruit he would be glad to cross a young tree with this. Budding is a kind of crossing.

"This work should be done in the spring, although it may be done in the fall. But the spring is a more limber time with Nature. Sap is begining to flow; life is new and fresh again; all the plant world is ready to start up and do something. Then, too, the bark of trees should be in as flexible a condition as possible. The two things really necessary for the work are mature buds and bark easy to peel.

"Buds should be taken from the very strongest and best twigs of the last year's growth. The little seedlings in which the buds are to be inserted should be one year old. These are called the stock.

"This is the manner of inserting the bud: first make a T-shaped cut in the bark of the stock. This cut should be made on the north side of the little tree, because it will thus be more sheltered from the sun's rays. The cut should not be far from the ground on the main trunk, although it may be at the base of strong shoots. But make it in the former position for these yearlings. Then loosen the bark with the flat handle of a regular budding knife. Not many boys and girls own such knives. Some of you have scalpels. The handles of these are flat enough to use. Again, you could easily whittle a piece of wood thin and flat enough for this work.

"The next question is how to sever a bud from its parent shoot. Suppose you have chosen a nice full bud. About a quarter of an inch below the base of the bud start cutting into the wood. Run the knife up to about one quarter of an inch above the bud. Do not cut out through to the surface, but rather from the upper surface cut the bark loose and peel this carefully down until you can see the under surface of the bud. You still have the wood attached to the twig at its upper edge. If as you look at the under surface of the bud you see that it is hollow, throw the whole thing away. If it has fibres then it is all right. The proper layer is left to reunite with the stock. Now the bud and peeled-off bark may be inserted in the T-shaped slip. Bind the bud in place with raffia. Do this raffia bandaging both above and below the slit.

"In about ten days the bandage may come off, for the knitting of fibres is well under way. Now the top of the little tree should be cut right back to about two inches above the bud, because you wish all the growth to go to the bud. This is the part of promise to the tree. All its hope lies in this new bud.

"The best method of increase is that of grafting. A graft or scion, which is a shoot with two or more buds on it of last year's growth, is inserted on the stem of another plant called the stock.

"By means of this process of grafting, trees bearing poor fruit are made to bear good fruit. Wild fruit trees are brought under cultivation, and a given tree may bear several varieties of its given fruit. For example, I have in mind a tree, the marvel of my childhood, which bore big sour apples, beautiful Gravensteins, and a good quality of Baldwins. This sort of experimenting with trees is not only as good as a puzzle, but is of great value. To make a wild apple tree with its gnarly, little sour apples into a really truly, well-behaved tree bearing good fruit is worth while, is it not? Grafting is not only a method of improvement but of taming stock, which is after all improvement.

"There are certain necessary precautions to take in this operation of grafting; for this, like budding, is a surgical operation.

"In all woody branches the outer layer is the bark; next comes the green layer. Between this layer and the real wood is a mass of fibres which go by the name of 'cambium'. The cambium layer of the stock and the scion must be one against the other in grafting in order that the sap may flow freely as before. This layer of cambium might be likened to our blood system. The two cambium layers must be pressed closely together so that the operation may be successful. And finally no cut surface should be left exposed to the air. It is air, you know, that plays havoc with flesh wounds. More and more we see that tree doctors have a work something like our own physicians.

"Grafting is usually done in the spring—in March or April—about the time sap begins to flow. The grafts or scions may be cut before this. Choose the tree from which you wish to take a scion. You choose it because of its fine-flavoured, sound fruit. Perhaps the fruit is especially large, too. Size of fruit, however, does not denote fine fruit. I once had an apple that weighed a pound. It was a beauty, fair to look upon. But what a tasteless, pithy piece of fruit it was. Appearances in fruit are often deceitful. The scions were to be of the last year's growth with two or more buds. The shoots should be clean, healthy and vigorous. You must transmit no disease along with the scions. These may be cut off in January or February, and stuck into the soil for about four to six inches. Keep away from direct sunlight. The buds of scions cut at this time are dormant as they should be.

"Grafting is named according to the manner in which the scion is put into the stock. There is whip grafting, and cleft grafting. Whip grafting is sometimes spoken of as tongue grafting.

"This latter method is accomplished in this way: Suppose you have a scion in your hand. Cut across the end of it diagonally. Use a sharp knife for this, and make a clean cut, as I now cut across this twig. About two-thirds of the distance back from the narrow or more pointed end of the cut make a vertical cut of about an inch right up into the scion. Cut the stock in a similar way. Then insert the tongue of the stock into the slit of the scion. Press these together carefully. Bind with raffia. Whenever this work is done outdoors, as it would be in the case of any of you who try this experiment, the union must be sealed over. As official documents are sealed with wax, so this union is legally sealed in wax. One can buy a regular grafting wax. Sometimes people mix clay and grease together. That is simple, but pretty sticky sounding.

"Realgrafting wax is made this way: To two parts of beeswax, add four of resin. Melt these together with one pound of tallow or linseed oil. When all are melted together, pour into cold water. Pull like molasses candy until it is light coloured. One's fingers should be greased to apply this wax properly.

"Cleft grafting is almost described by its name. A cleft or cut is made in the stock after the stem has been neatly cut across. The cleft is a vertical cut of about an inch in length. This is made through the centre of the stock. The scion is made to fit down into this, so naturally it is cut like a wedge. But there should be cuts made on both sides of the scion diagonally to form this wedge. So two cut surfaces of cambium are laid bare to fit against two similar surfaces of the stock. If the stock is several times thicker than the graft or scion, there should be two of these latter inserted. Place one at either end of the cleft. Bind and wax.

"If the stock is the same thickness as the graft then these two fit perfectly one into the other.

"This is only a little bit about grafting; but I trust this is enough to get you all interested in this work.

"'Is grafting really necessary?' I heard Albert whisper a while ago. It does seem like a great deal of work. The trouble with starting fruit from seed and expecting to get good results lies in this point: Fruit trees seems to lose in their development from seed the ability to produce fruit as fine as the parent stock; and so grafting becomes a necessity. Strange that this should be so, but it is.

"Start with a peach stone or seed. It came from a fine tree; the fruit was luscious. And yet the little seedling which comes from that very stone as a rule must be grafted to bear fruit of equally fine flavour as that of the original peach. Fruit trees have a tendency to revert to old wild poor forms. And so we must save them and help them.

"If any of you should start a little orchard he would wish to know how far apart the trees should be. Apple trees should be set thirty to forty feet apart each way; pear trees twenty to thirty feet each way; plums and peaches sixteen to twenty feet each way. Trees need room in which to spread out and develop; hence the distance given them. I am glad that Myron has made a start on small fruits. His strawberries were a success. I'd like to think that next season each of you was to have in his garden, vegetables, flowers, one small fruit and one of the larger ones, such as a seedling apple or peach."

VI

GARDEN OPERATIONS

"I suppose the talk to-day will seem to you all merely a repetition of things you already know. Beginnings, however, are most important. Results often take care of themselves, but beginnings never do. Gardens started wrong always go wrong; that is, unless one tears up one's work and begins over again.

"The first thing in garden making is the selection of a spot. Some of us are saved that trouble, since we have no choice; or like Josephine, have nothing at all in the way of space. Without a choice, it means simply doing the best one can with conditions. With space limited it resolves itself into no garden, or a box garden. Surely a box garden is better than nothing at all. At least, Josephine felt this to be true, and proved that parsley grows (with care) as well in a box as in the garden. I claim that everyone may have something of a garden if he be willing to take what comes to hand.

"But we will now suppose that it is possible to really choose just the right site for the garden. What shall be chosen? The greatest determining factor is the sun. No one would have a north corner, unless it were absolutely forced upon him; because, while north corners do for ferns, certain wild flowers, and begonias, they are of little use as spots for a general garden.

"If possible, choose the ideal spot—a southern exposure. Here the sun lies warm all day long. When the garden is thus located the rows of vegetables and flowers should run north and south. Thus placed, the plants receive the sun's rays all the morning on the eastern side, and all the afternoon on the western side. One ought not to have any lopsided plants with such an arrangement.

"Suppose the garden faces southeast. In this case the western sun is out of the problem. In order to get the best distribution of sunlight run the rows northwest and southeast.

"The idea is to get the most sunlight as evenly distributed as possible for the longest period of time. From the lopsided growth of window plants it is easy enough to see the effect on plants of poorly distributed light. So if you use a little diagram remembering that you wish the sun to shine part of the day on one side of the plants and part on the other, you can juggle out any situation. The southern exposure gives the ideal case because the sun gives half time nearly to each side. A northern exposure may mean an almost entire cut-off from sunlight; while northeastern and southwestern places always get uneven distribution of sun's rays, no matter how carefully this is planned.

"The garden, if possible, should be planned out on paper. The plan is a great help when the real planting time comes. It saves time and unnecessary buying of seed. Last winter we drew some plans to a scale. Peter, Philip, and Myron did this work in fine shape. They offer to take groups of you girls and show you how to do it; so whenever you are ready for this, the boys are ready, too. Sometimes we do change our plans some, anyway a change is easily made when a plan has been drawn as a basis.

"New garden spots are likely to be found in two conditions: they are covered either with turf or with rubbish. In large garden areas the ground is ploughed and the sod turned under; but in small gardens remove the sod. How to take off the sod in the best manner is the next question. Stake and line off the garden spot. The line gives an accurate and straight course to follow. Cut the edges with the spade all along the line. If the area is a small one, say four feet by eighteen or twenty, this is an easy matter. Such a narrow strip may be marked off like a checkerboard, the sod cut through with the spade, and easily removed. This could be done in two long strips cut lengthwise of the strip. When the turf is cut through, roll it right up like a roll of carpet.

"But suppose the garden plot is large. Then divide this up into strips a foot wide and take off the sod as before. What shall be done with the sod? Do not throw it away for it is full of richness, although not quite in available form. So pack the sod grass side down one square on another. Leave it to rot and to weather. When rotted it makes a fine fertilizer. Such a pile of rotting vegetable matter is called a compost pile. All through the summer add any old green vegetable matter to this. In the fall put the autumn leaves on. A fine lot of goodness is being fixed for another season.

"The girls, I suppose, think this is a wretched heap to have in the corner of a garden. So it is. But it is possible to screen it. Plant before the space allotted to this, castor beans, tall cannas or sunflowers. Perhaps the castor beans would be the best of all. Sunflowers get brown and straggly looking before the season is past its prime.

"Even when the garden is large enough to plough, I would pick out the largest pieces of sod rather than have them turned under. Go over the ploughed space, pick out the pieces of sod, shake them well and pack them up in a compost heap.

"What is to be done with the rubbish often found on new garden sites? If this be only weeds and other vegetable matter it may be very easily burned on the garden spot. But suppose it is a grand collection of tin cans, bottles and such things as cannot be burned? What can we do with them? Cities have public dumps where lots are to be filled in. All such trash may go to these. Oftentimes it is possible to find suitable places in the country for dumping. But do not dump where the rubbish is to be unsightly for others as it has been for yourself; far better have a dump heap on your own land and screen this as the compost heap was to be shielded from view. We take the wrong point of view if we dump rubbish anywhere, for the sake of getting rid of it. You remember your plan is to help make a more beautiful village.

"How must the small garden be spaded? A method called trenching, is good because it is so thorough. Here is a diagram George has made. Just get your heads around this, and I'll explain it.

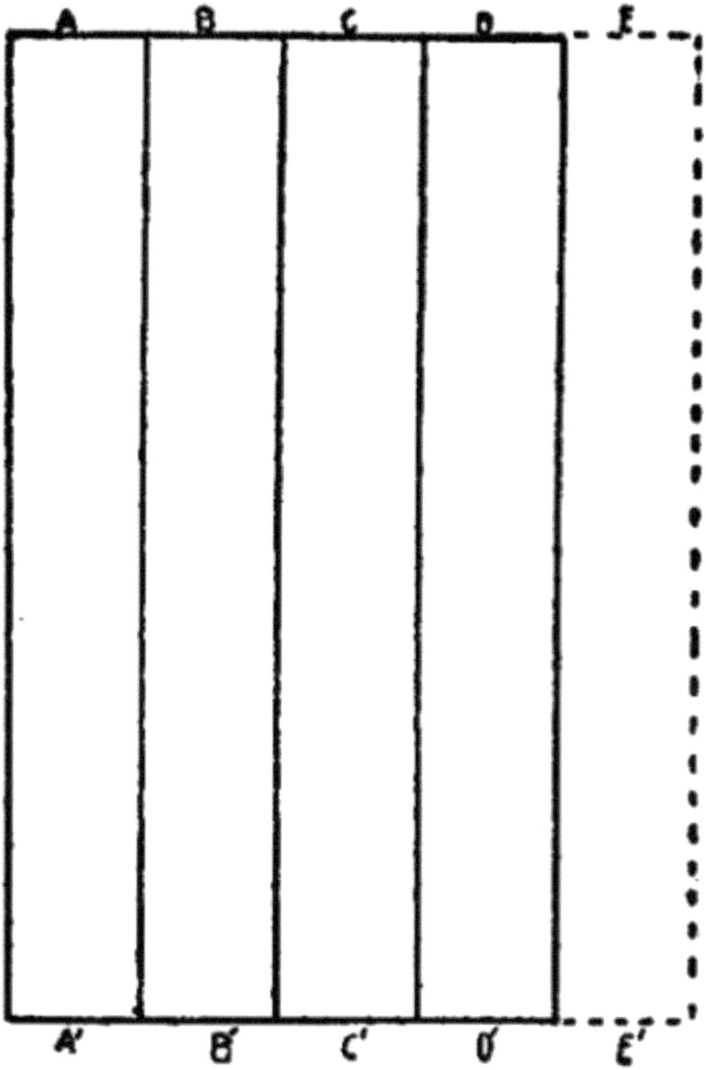

From this plan see the scheme of trenching. Top soil from AA' is carted to EE'. Then the top soil from BB' goes into AA'. Continue this method and see that the soil on EE' finally goes into trench

DD'. So all the top soil in this given area is worked over and is still kept on top.

"This rectangle is supposed to be the plot which needs digging. Line it off into strips one foot wide. Have your wheelbarrow right beside AA'. Dig one foot of top soil out of strip A' along all its length. Put this into the barrow and dump it into the strip marked EE' outside of the garden proper. Do the same thing to strip BB', only throw the soil into trench AA'. The top soil from CC' goes into BB'; that of DD' into CC'. Now the soil that was dumped outside the garden upon the strip EE' of course is already to go right into trench DD'.

"The value of this work is to get the soil of the bed entirely worked over. Most people dig but poorly. Digging is hard work; so a boy digs a little here, and a little there, throughout the seed bed and thinks the work is all done. It is really done when the above method is used. And after all we have said about the necessity for airing soil, and the need of stirring things up so that the good bacteria may do their work, I know you will all see the point immediately.

"Mere spading of the ground is not sufficient. The soil is still left in lumps. Always as one spades one should break up the big lumps. But even so the ground is in no shape for planting. Ground must be very fine indeed to plant in, because seeds can get very close indeed to fine particles of soil. But the large lumps leave large spaces which no tiny root hair can penetrate. A seed is left stranded in a perfect waste when planted in chunks of soil. A baby surrounded with great pieces of beefsteak would starve. A seed among large lumps of soil is in a similar situation. The spade never can do this work of pulverizing soil. But the rake can. That's the value of the rake. It is a great lump breaker, but will not do for large lumps. If the soil still has large lumps in it take the hoe.

"Many people handle the hoe awkwardly. Get up, Jay, and show us just how to hold it! Walk along as you hoe, drawing the hoe toward you. The chief work of this implement is to rid the soil of weeds and stir up the top surface. It is used in summer to form that mulch of dust so valuable in retaining moisture in the soil. I often see boys hoe as if they were going to chop into atoms everything around. Hoeing should never be such vigorous exercise as that. Spading is vigorous, hard work, but not hoeing and raking.

"After lumps are broken use the rake to make the bed fine and smooth. Now the great piece of work is done. To be sure I have said nothing of fertilizing. The kind and amount of fertilizer depends on the kind of soil. Well-rotted manure being the best all-around fertilizer, we will say that we have spaded that into the seed bed after the trenching operation is over.

"Now the plan made on paper comes into practical use, and garden stakes, cord and a means of measuring are the things necessary to have on hand. Jay and Albert have made their garden stakes one foot in length. They will serve as a good rule in furrow making. On their hoe handles Jack and Elizabeth have marked two feet off into inches. This is another scheme for measuring. George has a pole four feet long which he uses. This has inches marked on one foot of its length. Katharine has a seventy-five foot tape measure. And Leston and Helena have made this tool I have here in my hand. It looks like a wooden toothed rake with its teeth eight inches apart. This dragged over the surface of a nice, fine garden bed marks off furrows. It makes the most regular furrows you ever saw because it cannot help itself. Miriam used a board last summer. She laid this across her seed bed, kneeling on it, then she drew a dibber along the board's straight edge, pressing firmly into the soil with the dibber. This also made a good straight furrow.

"Peter and Philip always use a line and two stout garden stakes. Their hoes do the rest.

"We usually think of furrows, or drills, as they really should be called when little soil is removed, as being about a half inch or even less in width. Sometimes certain seed, beans and peas, for example, are placed in double rows in a wide drill.

"I think you all understand hill making. Then you remember how we planted certain seeds broadcast, as grass and poppy seeds. Remember that seeds thus sown need only a dusting of soil over them.

"But in general, drill sowing for both vegetables and flower seeds is the most satisfactory method.

"Most boys and girls sow seeds too thickly. The seedlings as they come up are too crowded for proper amounts of sunlight, air and food. You have seen lettuce seedlings crowded together growing small and weak. Why? Lack of light and air, lack of moisture and food are the reasons for this. Thin out pretty severely. Wait, of course, until the seedlings are an inch or more high. Then look over the little plants and gently take out the weakest and smallest specimens. Press the soil firmly about those which remain. If the first planting has been very thick have two times of thinning. It is a bit easier on those seedlings remaining if too many comrades do not go at once.

Jack's Rake Handle as a Measuring Stick

Albert Sowing Large Seeds Singly

Photographs by Edward Mahoney

"Of course, some of these seedlings may be transplanted. They should be about two inches above ground for this purpose. Lettuce, cabbage and peppers transplant beautifully; so do asters. I would not try to transplant beets, radish or turnips. The reason is that these plants have long tap roots. Usually a portion of the root is left in the ground and the transplanted seedling has an injured root. So you either lose it, or it does poorly.

"Beets may be allowed to grow thickly for a time. Then when the thinning is done, the tender beet tops may be used for greens.

"Transplanting is a delicate operation. A trowel or a thin garden marker, a can of water and dibber are the necessary tools for the business. A cloudy day is a good thing to have on hand, also. If this is impossible, place the sun behind a cloud. The little seedling should be taken up with great care from its old home. A little soil should come with the roots. This gives the little plant a home feeling in its new quarters. The thin stick is often better to use than the trowel. If the soil is watered a bit about the small plant, one is far more likely to get the soil up with the roots.

"Now make the hole in the ground with the dibber just where you wish. A motion, like that of a revolving top, is the one to use in working the dibber. Water the hole. Drop a little soil in the bottom of the hole. You see the dibber leaves an awkward little peak there at the bottom of the hole. Water lodges there and stays. The tiny rootlets do not quite reach into the bottom of the hole, and perhaps dangle in the water and begin to decay. A little soil dropped in prevents all this. Now a little plant goes in. Do not place it too low, nor too high in the hole. Have the roots uncramped. Drop soil in gently and finally firm it all with both hands.

"The sun must not shine too hotly for the first few days on these little plants in their new home. They are not yet used to their surroundings and must be coddled a bit if they are to do well.

"The remaining garden operations are weeding and constant cultivation. A part of the work in the flower garden is close picking, if constant bloom is wished.

"I have said nothing about how to plant different seeds because each of you had tables to cover all of that.

"The object of this talk is to impress upon you the necessity for careful preparation. Well-prepared soil, carefully handled tools and plants are ways to success.

"Good tools, good seed, good hard work make for results such as will satisfy your highest hopes. But it is not the result only that is worth the struggle; the knowledge and the power are the greater glories."

VII

COMMON WEEDS

What a delight it would be if we could garden without weeds. But that is well-nigh impossible. For these rascals, the weeds, are such persistent fellows, so clever in their devices for getting over the surface of the earth, so able to live where nothing else in the plant world can live, that it is a discouraging matter to attempt to exterminate them. They always seem to me like pushing sort of people trying to live among those who do not want them. Then, too, they crowd the better class of inhabitants out.

"There are a certain number of plants which we always looked upon as weeds, such as burdock and wild carrot, for example. But if a beautiful garden plant should persist in living and spreading itself over our vegetable garden, then that, too, would become a weed. Over across the sea in England the poppy grows wild in the fields. It looks very beautiful to the traveler, because it makes lovely red splashes of colour through the field. But I doubt very much if it looks really attractive to the farmer. These things depend largely, do they not, upon one's point of view?

"Even a question like weeds we have no right to look at from one point of view only. The good points of weeds do not balance up the bad points; but it is well to give even weeds their due. Rid the world of weeds and unless these despoiled spots were cultivated, think of the great waste places there would be over the earth's surface. The weeds shade the ground thus preventing too great surface evaporation. Then the weeds are a signal to farmers and all gardeners to get busy. We people of the world are lazy, just naturally so, and perhaps if there were no weeds we might cultivate the soil too little. Years ago certain weeds were much used in medicine. This is more or less true, to-day. The dandelion with its bitter secretion was good, it was believed, for the liver, a sort of spring tonic. The Department of Agriculture has printed a pamphlet on 'Weeds Used in Medicine' (Farmers' Bulletin, No. 188). Jack and Jay each sent for a copy last spring. You all might start a garden library with these pamphlets for a basis. They are sent to you free and are invaluable in your work. Get together all the helps you can on the subject you are studying. Boys and girls receive free so much in the present day that it seems a shame not to make use of these things. The boys have written to the Department of Agriculture and each month it sends to the club a list of the publications sent out or reprinted during the previous month. You girls might follow this good example set you by the boys.

"Well, we have wandered a bit from the subject in hand. Weeds are again discouraging because they have such facilities for travel. Talk about flying machines—weeds are centuries ahead of men along these lines. Look at a milkweed seed; it is a complete flying apparatus. With its perfect ballast it flies beautifully along over field and river ready to alight in proper seed style, end down.

"There is a piece of mechanism in the end of each burdock seed that seems to make travel possible, and dissemination sure. Never was fish hook more cleverly made than this hook of the bur seed. It catches on to your clothing and travels until you feel its pull. Then you pick it off and cast it aside. So it goes. It sticks to the furry and hairy coats of animals and again is carried along.

"Did you ever observe the seed of wild carrot? It, too, is arranged with clinging points all around and about its seed. If you should give just a little attention to the subject of the means of distribution of wild seeds you would have a greater respect for the ways and means of Nature.

"Here is another discouraging side to the weed question. Weeds produce so many, many seeds! Look at a single stalk of plantain. This stalk does not stand for one seed capsule, but all up and down the stalk are the seeds; again, not one seed here and one there, but each capsule or seed case holding many seeds. When these become ripe, then the top of the capsule comes off just like the cover of a box, or the top of a salt cellar, and the seeds are sent out. It would not be a useless thing to count sometimes the number of seeds on one plantain stalk, and thus gain an idea of the tremendous possibilities of increase which the weeds have.

"A lad I once knew counted the number of seeds in a milkweed pod which he had, and found very nearly two hundred. I do not remember the exact number. It was between one hundred and ninety-five and two hundred. Think of one pod scattering that number of seeds! Think again of the number of pods on one milkweed plant! It is staggering, is it not? To be sure we can remember the parable of the sower and have some hope, for some seed may fall on soil in which they will never come to maturity.

"Weeds, like the wild morning glory, form new plants not from their seed only, but from their travelling, trailing branches.

"If, then, the chances are so good for renewal of weeds, what is the plan of campaign which we should follow? Once a German gentleman who loved and cultivated roses was asked how to get rid of rose bugs. 'Kill them,' he said. 'Pick them off by hand and kill them by foot is the sure method!' he continued.

"So, to get rid of weeds, just destroy them. Persistently and constantly weed them out and cultivate the soil. Clean cultivation is the only sort for good crops and freedom from weeds.

"Weeds, as flowers, drop in the three classes of annuals, biennials and perennials. Any annual is easy enough to hold down. Just pull such weeds up. Some merely cut the weed off at the surface of the ground, but it is a better way to be rid of the thing entirely. And should you not be quite sure of the kind of weed, then pulling up is the only really safe plan. For if the weed happened to be a perennial, leaving the root in the ground would be the worst possible thing to do.

"The greatest business of all annuals is to form seed. Now I know you wish to say that this is the business of all plants. It is. But with annuals there is only one chance to produce seed. That chance is the one short year of their lives, and this is doubtless the reason why these chaps work so hard at seed forming, and produce so many seed. Therefore, the thing evidently to be done is to make it impossible for annuals to form seed.

"The biennials and perennials must have further treatment than just that of preventing seed formation. The underground part of such weeds must be destroyed. For these live in the ground ready to come up again. Biennials may be killed out by deep hoeing. Get rid of all the young plants, keep at the older ones with the hoe and prevent seed formation, too. Biennials are found most abundantly in waste places along woodsides and where the soil for a long time has been left undisturbed.

"Perennials need about the same treatment as biennials. But even greater persistency should be exercised in destroying the underground portion. For these underground plants produce new plants as surely as seeds do. The bindweed has a creeping root, wild garlic has a bulb, and such forms are always producing new forms underground while the seed above the ground is able to do the same thing.

"Ploughing helps destroy perennials, as the roots are exposed to direct sunlight and so destroyed. Another method of treatment is that of cutting off the top down to the root and putting salt on the freshly cut root tap. Then again these roots may be starved out by never allowing the top or leafy part to form. You will remember that it is the leaf which makes the food. And if there is no food then there will be none to store away in the root for new root formation. Some farmers smother roots. This is done by planting such crops as hemp, clover or cowpeas. These crops choke out the weeds. They cover the ground very completely, and so the weeds have less of a chance.

"I give the following table of a few very common weeds in order that you may know just how to handle them.

"I must speak especially about snapdragon or butter and eggs. It came to our country as a garden flower. It has spread and spread, partly by its seeds and partly by its root stalks, which are creeping ones, and now it is a perennial weed. For since it has become a nuisance it must be classed as a weed. As it spreads along it tends to force out other plants.

"This weed, like the wild carrot, is really very lovely. Could such weeds be properly held down in small garden areas they would be very ornamental. I saw a little flower garden once, quite beautiful, with two small clumps, one of wild mustard and one of field daisy, among the other flowers.

COMMON NAME	CLASS	SEED TIME	COLOUR OF FLOWER
Burdock	Biennial	Aug.-Oct.	Purple
Bur-marigold or beggar ticks	Annual	July-Nov.	Yellow
Canada thistle	Perennial	" "	"
Chickweed	Annual	Mar.-July	White
Cocklebur	"	July-Oct.	Green
Dandelion	Perennial	May-Oct.	Yellow
English bindweed or morning glory	"	Aug.-Oct.	White
Moth Mullein	Biennial	July-Nov.	Yellow
Narrow-leaved stickseed or beggar tick	Annual	July-Oct.	Blue
Ox-eye daisy or white daisy	Perennial	Aug.-Oct.	White
Pigweed	Annual	Aug.-Nov.	Green
Prickly lettuce, milkweed	"	July-Nov.	Yellow
Purslane, pursley	"	June-Dec.	"
Rib-grass, plantain	Biennial	July-Nov.	White
Ragweed, Roman wormwood	Annual	Aug.-Nov.	Yellow
Russian thistle	"	" "	Purplish
Smartweed	Perennial	Aug.-Sept.	Pink
Sorrel	Perennial	June-Nov.	Red
Wild carrot, Queen Anne's lace	Biennial	July-Nov.	White

Wild garlic, onion	Perennial	July-Oct.	"
Wild mustard	Annual	June-Oct.	Yellow
Wild parsnip	Biennial	July-Oct.	"
Yellow daisy, ox-eye daisy, brown-eyed Susan	"	July-Sept.	"
Yellow dock	Perennial	Aug.-Oct.	Green

"The seeds of the wild mustard, like those of the plantain and other weeds, get in with the grain seed and so cause constant trouble. Farmers feel that such weeds must be thoroughly gotten out of the fields.

"It is not our own native weeds which are so troublesome but the foreign ones. Most of our worst weeds are foreigners. They have come to this country as stowaways from across the seas. They have fought for centuries and can keep the fight up over here.

"I am not going to give you a description of each weed we have. This table, a copy of which is for each of you, will be, I think, of true help. The study of weeds is something quite by itself. It is for you to help prevent the seeding of weeds everywhere. Do not carelessly scatter seeds. Keep your own garden plots free from these pests by clean and careful cultivation. Remember, too, the value of cover crops.

"There is another pest to fight. This pest is of the animal kingdom and not of the plant kingdom. Next Friday our talk is on animal pests, and how to destroy them."

VIII

GARDEN PESTS

If we could garden without any interference from the pests which attack plants, then indeed gardening would be a simple matter. But all the time we must watch out for these little foes—little in size, but tremendous in the havoc they make.

"As human illness may often be prevented by healthful conditions, so pests may be kept away by strict garden cleanliness. Heaps of waste are lodging places for the breeding of insects. I do not think a compost pile will do the harm, but unkempt, uncared-for spots seem to invite trouble.

"There are certain helps to keeping pests down. The constant stirring up of the soil by earthworms is an aid in keeping the soil open to air and water. Many of our common birds feed upon insects. The sparrows, robins, chickadees, meadow larks and orioles are all examples of birds who help in this way. Some insects feed on other and harmful insects. Some kinds of ladybugs do this good deed. The ichneumon-fly helps too. And toads are wonders in the number of insects they can consume at one meal. The toad deserves very kind treatment from all of us.

"Each girl and boy gardener should try to make her or his garden into a place attractive to birds and toads. A good birdhouse, grain sprinkled about in early spring, a water-place, are invitations for birds to stay a while in your garden. If you wish toads, fix things up for them too. During a hot summer day a toad likes to rest in the shade. By night he is ready to go forth to eat but not to kill, since toads prefer live food. How can one "fix up" for toads? Well, one thing to do is to prepare a retreat, quiet, dark and damp. A few stones of some size underneath the shade of a shrub with perhaps a carpeting of damp leaves, would appear very fine to a toad.

"Suppose a certain crop in your garden has had an insect pest. Do not plant this same crop next year, for it would doubtless have the same pest. Don't let the soil get full of insect troubles; therefore, keep the soil open and aired and study it well.

"There are two general classes of insects known by the way they do their work. One kind gnaws at the plant really taking pieces of it into its system. This kind of insect has a mouth fitted to do this work. Grasshoppers and caterpillars are of this sort. The other kind sucks the juices from a plant. This, in some ways, is the worst sort. Plant lice belong here, as do mosquitoes, which prey on us. All the scale insects fasten themselves on plants, and suck out the life of the plants.

"Now can we fight these chaps? The gnawing fellows may be caught with poison sprayed upon plants, which they take into their bodies with the plant. The Bordeaux mixture which Peter used is a poison sprayed upon plants for this purpose. So, too, is Paris green.

"In the other case the only thing is to attack the insect direct. So certain insecticides, as they are called, are sprayed on the plant to fall upon the insect. They do a deadly work of attacking, in one way or another, the body of the insect. The kerosene emulsion made by the girls for their infested house plants worked this same way. Tobacco water and tobacco dust sprinkled on act in similar manner.

"Lime, soot, and sand are other means of blocking and choking off insects.

"Sometimes we are much troubled with underground insects at work. You have seen a garden covered with ant hills. Here is a remedy, but one of which you must be careful.

"Carbon bisulphid comes in little tin cans. It is a liquid of a vile smell, something like onions and rotten eggs mixed. The girls' noses are going up sky-high now. But it does the work of ant killing. You must be careful in handling this. It has a horrid explosive habit. Pour about a teaspoonful down an ant hole. Do not use a good silver spoon from the dining room. Get an old spoon, or buy a tin one. For you will never use it again except it be for carbon-bisulphid work. After this liquid has been poured down the hole, place a bit of a chip over it, for there may be a slight volcanic action underground. It is well to do this on a damp, cloudy day when all the ants are at home.

"Remember this stuff is not to be fooled with, as it is poisonous and also takes fire readily. Never open the can inside by a fire, in too great a heat, or near a lighted match. Invite your fathers to help in this. By no means do anything silly. Keep the can closed except when pouring out a teaspoonful.

"This question is constantly being asked, 'How can I tell what insect is doing the destructive work?' Well, you can tell partly by the work done, and partly by seeing the insect itself. This latter thing is not always so easy to accomplish. I had cutworms one season and never saw one. I saw only the work done. If stalks of tender plants are cut clean off be pretty sure the cutworm is abroad. What does he look like? Well, that is a hard question because his family is a large one. Should you see sometime a grayish striped caterpillar, you may know it is a cutworm. But because of its habit of resting in the ground during the day and working by night, it is difficult to catch sight of one. The cutworm is around early in the season ready to cut the flower stalks of the hyacinths. When the peas come on a bit later, he is ready for them. A very good way to block him off is to put paper collars, or tin ones, about the plants. These collars should be about an inch away from the plant.

"Of course, plant lice are more common. Those we see are often green in colour. But they may be red, yellow or brown. Kerosene emulsion is the medicine for plant lice. Lice are easy enough to find since they are always clinging to their host. As sucking insects they have to cling close to a plant for food, and one is pretty sure to find them. But the biting insects do their work, and then go hide. That makes them much more difficult to deal with.

"Rose slugs do great damage to the rose bushes. They eat out the body of the leaves, so that just the veining is left. They are soft-bodied, green above and yellow below. Since they are eating insects Paris green will kill them. But the kerosene emulsion penetrates their soft bodies; so this also may be used.

"A beetle, the striped beetle, attacks young melons and squash leaves. It eats the leaf by riddling out holes in it. This beetle, as its name implies, is striped. The back is black with yellow stripes running lengthwise. White hellebore powder kills these pests. Ask the druggist for five cents' worth and you will have a great plenty for any of your gardens. It, too, is a poison. This poison is also good to use for the caterpillars that eat many of our garden plants. Make a circle four inches from the stalk of an infested plant and sprinkle the powder in this. Evening time is good for this, because the dew moistens the powder just enough to make it a nuisance to the insect.

"Then there are the slugs, which are garden pests. The slug will devour almost any garden plant, whether it be a flower or a vegetable. They lay lots of eggs in old rubbish heaps. Do you see the good of cleaning up rubbish? The slugs do more harm in the garden than almost any other single insect pest. You can discover them in the following way. There is a trick for bringing them to the surface of the ground in the day time. You see they rest during the day below ground. So just water the soil in which the slugs are supposed to be. How are you to know where they are? They are quite likely to hide near the plants they are feeding on. So water the ground with some nice clean lime water. This will disturb them, and up they'll poke to see what the matter is.

"Beside these most common of pests already mentioned, pests which attack many kinds of plants, there are special pests for special plants. Discouraging, is it not? Beans have pests of their own; so have potatoes and cabbages, as George well knows. In fact, the vegetable garden has many inhabitants. In the flower garden lice are very bothersome, the cutworm and the slug have a good time there, too, and ants often get very numerous as the season advances. But for real discouraging insect troubles the vegetable garden takes the prize. If we were going into fruit to any extent, perhaps the vegetable garden would have to resign in favour of the fruit garden.

"A common pest in the vegetable garden is the tomato worm. This is a large yellowish or greenish striped worm. Its work is to eat into the young fruit.

"A great, light green caterpillar is found on celery. This caterpillar may be told by the black bands, one on each ring or segment of its body.

"The squash bug may be told by its brown body, which is long and slender, and by the disagreeable odour from it when killed. The potato bug is another fellow to look out for. It is a beetle with yellow and black stripes down its crusty back. The little green cabbage worm is a perfect nuisance. It is a small caterpillar and smaller than the tomato worm. These are perhaps the most common of garden pests by name. It might be well to take up the common vegetables and flowers mentioning the pests which prey on each one.

"Let us take the vegetables first. None of us have grown asparagus yet; but it will be well to know about this vegetable. There is a beetle which may trouble asparagus plants. It is red with markings of black. The grub of this beetle is dark green. Look out for the asparagus beetle during April and May, for these are the months when it appears. The eggs are laid on young shoots of the plant. Such shoots should be cut right off. After the cutting season is over the plants should be sprayed. This may be done in August. Very dilute Bordeaux mixture or Paris green may be used for a spray.

"Next in the alphabet come beans. The most common trouble that beans have is one called anthracnose. That staggering word means that the leaves become covered with spots which are round with purple borders. Again, a spray of Bordeaux mixture should be used. The plants should be sprayed until the pods form. Look for this trouble in July.

"Beets are prone to leaf spots. As soon as such spots appear, the plants should be sprayed with the Bordeaux mixture. Every two weeks give the plants about three sprayings.

"The cabbage worm I have spoken of. This worm works all summer. Cabbages, if neglected, become literally alive with the little caterpillars. They eat and eat the foliage, riddling it completely. They eat into the heads so that the cabbage plant is completely spoiled. George treated his with pyrethrum powder. This he mixed with five times its bulk of dust. It was then dusted or shaken over the cabbage plant. A very good thing to do before trouble begins is to dust the soil and tender plants with lime. After the plants have begun to head use hellebore powder.

"Lice appear on cauliflower. The kerosene emulsion which we use on our indoor plants is all right for this work, too. The lice appear on the foliage in great white masses. They suck the life and goodness from the plant. They come all through the summer at any time. Whale oil soap is another good spray to use. Peter has typewritten receipts for these sprays which you may have at the close of this talk. Sometimes the root of the cauliflower is attacked. Little white maggots mine or burrow through the root. They are quite likely to begin their bad work in June or July. That rather dangerous carbon bisulphid is the medicine for this trouble. Make a hole in the soil as you did when treating the ant. Do not make this too near the plant. I should say six inches away would be about right. Pour a teaspoonful of the poison into the hole and it will take care of itself. Cover the hole over as you would in the case of the ant. When cauliflower plants begin to look sickly pull one up. If it is full of maggots that is easy to determine. But it may be that you will find great lumps or knots on the root. Since these knots appear during the same months as the maggots, you can only be sure of the real cause by pulling up a plant. If these knots are on the root, then you have a very serious trouble to contend with. So serious is the club root condition that the only safe thing to do is to pull up and completely destroy the diseased plants. Dig the soil up after this. Then lime it. Put a lot of lime on, not just a dusting over the surface of the soil. This represents soil that is in trouble, so do not plant cauliflower here again, or its coarser cousin, the cabbage.

"Sometimes a little red or orange and black bug appears. This is called the harlequin bug from its fantastic appearance. This bug may come all summer long at any time. The whale oil soap spray is the one to use. Celery may be troubled with the light green caterpillar with the black bands before spoken of. This caterpillar arrives in August. It is not difficult at all to see, so many may be picked off just by hand. One may use Paris green as a spray.

"None of you had any trouble with corn being infested. But sometimes a worm, called the earworm, which is like the tomato worm, will appear during June and eat the tips of the young ears. A little Paris green sprinkled on the leaves, at their base will kill them.

"Cucumbers and melons, as I have before said, are prone to be preys of the cutworms, squash bugs, striped beetles, and lice or aphis. You know treatment for cutworms and lice. The squash bug may be destroyed by hand. Sometimes when bits of sticks are placed on the ground the bugs will crawl under them. Next morning a small harvest of bugs can be killed. The squash bug lays its eggs on the under surface of the plant's leaves. These leaves should be removed and burned. The striped beetle is kept off by the Bordeaux mixture spray. This beetle appears in June. A spraying during this month often prevents a blight of the leaves in July. This blight appears first as a spotting on the leaves, after which the leaves soon wither up.

"Onions, as well as radishes, are affected by maggots which will mine through the onion bulb as well as the stems of the young, tender plants. A solution made from carbolic soap and water is excellent with which to water the soil about the plants.

"Peas have green lice as melons and cucumbers do. The lice appear early in May and June, and are killed and kept down by the regulation treatment. Many times during the latter part of summer peas may become mildewed. You can tell this by a growth of white down on stem and leaves. Put some soap in the Bordeaux mixture and spray.

"From May to October potato bugs flourish. Paris green is the spray to use. In the start they may be hand picked. But do not let them get ahead of your hand. A very serious potato disease is that of scab. Scales appear on the potatoes themselves. To prevent this, uncut seed potatoes are soaked in poison. But this is not a work for you to do alone by any means.

"The squash bug naturally seeks out the squash vine. He should be treated as we said when we talked of the same bug and melons.

"Tomatoes have numerous troubles. The cutworm, the tomato worm, the horn worm, potato beetle and various blights may come to tomatoes. The horn worm is a large green worm named from the horn at one end of its body. It appears in midsummer. Such large worms usually may be hand picked. If you should see a tomato plant wilting for no reason at all, pull it up and burn it; it probably has an infectious trouble which is carried from one plant to another by insects. It is really an infectious disease.

"These are the most common vegetable garden pests and their remedies.

"As the girls know, the flower garden is not without pests, too. Plant lice are plenty enough. These may appear at one time or another during the entire year.

"Some plants become covered with a little red spider. It attacks the foliage and does great damage. This may be due to lack of moisture with house plants. I do not mean lack of watering, but a dry condition of the air of the room. Often just a spray of clear water is sufficient to rid the plant of the mites.

"Roses have more troubles than any one other flower. The rose bush may have lice or it may have a little green bug that jumps very quickly and so gets its name of leaf-hopper. Kerosene emulsion is good to use. Often slugs will feed upon the surface of the leaves. A dusting of lime over the leaves keeps these feeders away. There is a brown beetle called the rose chafer, which eats the flower itself. Hand picking is about the best weapon to employ against this enemy. A scale sometimes comes on the stems. This scale looks like a white crust. It is wise to spray such rose bushes with kerosene emulsion. And better still, if possible, cut off and burn such scale-encrusted parts.

"Cutworms bother the early bulbs and the violets, too. A great many of the larger pests may be hand picked. The lice should be sprayed.

"And for the remedies. The following will be the ones you will need the most:

KEROSENE EMULSION

Soap (Ivory)	¼ pound
Boiling water	2 quarts
Kerosene	4 quarts

"The soap should be shaved up and dissolved in the water. To this add the kerosene (of course not when the soap and water is on the stove) a little at a time. Beat it with an egg beater to be kept for that purpose; or shake it vigorously.

"For use against plant lice add to one cup of this emulsion 8 cups of water. For scale insects dilute with four cups of water.

BORDEAUX MIXTURE

I. Copper sulphate (blue vitriol)	1 lb.
Water	1 gal.

Dissolve the sulphate in the water.

II. Slaked lime	1 lb.
Water	1 gal.

Dissolve the lime in the water.

To one bucket (2½ gallons) of fresh water add four pints of the first solution. To another bucket of fresh water add six pints of the second solution.

Stir these together. Keep the rest of the solutions I. and II. for later mixing when it is needed.

WHALE OIL SOAP HOT WATER SOLUTION

Whale oil soap	1 pound
Hot water	6 quarts

This is the right dilution for plant lice but for scale insects it is too weak; for them use about two quarts of water to one pound of soap.

"The best way to apply liquid sprays in small gardens is to use a whisk broom. Just dip the little broom into the mixture needed and shake the brush over the plant. Then the hands need never come in contact with the poison. Careful children can use sprays without any trouble. Josephine has used kerosene emulsion in this fashion: she pours a little into a saucer, takes a bit of cheese cloth and dipping it into the emulsion wipes the lice off an infested part. Usually one application is enough. This sounds like a much more disagreeable task than it really is. A plant syringe may be used. But personally I like the hand method. Of course if there are lots of lice on many plants this would not be practical at all.

"It stands to reason that sick plants need medicines of some kind. Sometimes to be sure they need better living conditions. Often the soil is sour, water-logged, unaired and totally unfit for a self-respecting plant to live in. The whole thing resolves itself into a study of conditions, and a desire to help the plant have as comfortable a time as possible in life."

IX

VEGETABLE CULTURE

"As a rule, boys and girls choose to grow bush beans rather than pole beans. I cannot make up my mind whether or not this is from sheer laziness. In a city backyard the tall varieties might perhaps be a problem since it would be difficult to get poles. But these running beans can be trained along old fences and with little urging will run up the stalks of the tallest sunflowers. So that settles the pole question. There is an ornamental side to the bean question. Suppose you plant these tall beans at the extreme rear end of each vegetable row. Make arches with supple tree limbs, binding them over to form the arch. Train the beans over these. When one stands facing the garden, what a beautiful terminus these bean arches make.

"Beans like rich, warm, sandy soil. In order to assist the soil be sure to dig deeply, and work it over thoroughly for bean culture. It never does to plant beans before the world has warmed up from its spring chills. There is another advantage in early digging of soil. It brings to the surface eggs and larvae of insects. The birds eager for food will even follow the plough to pick from the soil these choice morsels. A little lime worked in with the soil is helpful in the cultivation of beans.

"Bush beans are planted in drills about eighteen inches apart, while the pole-bean rows should be three feet apart. The drills for the bush limas should be further apart than those for the other dwarf beans—say three feet. This amount of space gives opportunity for cultivation with the hoe. If the running beans climb too high just pinch off the growing extreme end, and this will hold back the upward growth.

"Among bush beans are the dwarf, snap or string beans, the wax beans, the bush limas, one variety of which is known as brittle beans. Among the pole beans are the pole limas, wax and scarlet runner. The scarlet runner is a beauty for decorative effects. The flowers are scarlet and are fine against an old fence. These are quite lovely in the flower garden. Where one wishes a vine, this is good to plant for one gets both a vegetable, bright flowers and a screen from the one plant. When planting beans put the bean in the soil edgewise with the eye down.

"Beets like rich, sandy loam, also. Fresh manure worked into the soil is fatal for beets, as it is for many another crop. But we will suppose that nothing is available but fresh manure. Some gardeners say to work this into the soil with great care and thoroughness. But even so, there is danger of a particle of it getting next to a tender beet root. The following can be done; Dig a trench about a foot deep, spread a thin layer of manure in this, cover it with soil, and plant above this. By the time the main root strikes down to the manure layer, there will be little harm done. Beets should not be transplanted. If the rows are one foot apart there is ample space for cultivation. Whenever the weather is really settled, then these seeds may be planted. Young beet tops make fine greens. Greater care should be taken in handling beets than usually is shown. When beets are to be boiled, if the tip of the root and the tops are cut off, the beet bleeds. This means a loss of good material. Pinching off such parts with the fingers and doing this not too closely to the beet itself is the proper method of handling. I throw this in for the benefit of our future cooks, the girls.

"There are big coarse members of the beet and cabbage families called the mangel wurzel and ruta baga. About here these are raised to feed to the cattle. They are a great addition to a cow's dinner.

"The cabbage family is a large one. There is the cabbage proper, then cauliflower, broccoli or a more hardy cauliflower, kale, Brussels sprouts and kohlrabi, a cabbage-turnip combination. George has worked out cabbage culture successfully. I refer to him for full particulars.

"Cauliflower is a kind of refined, high-toned cabbage relative. It needs a little richer soil than cabbage and cannot stand the frost. A frequent watering with manure water gives it the extra richness and water it really needs. The outer leaves must be bent over, as in the case of the young cabbage, in order to get the white head. The dwarf varieties are rather the best to plant.

"Kale is not quite so particular a cousin. It can stand frost. Rich soil is necessary, and early spring planting, because of slow maturing. It may be planted in September for early spring work.

"Brussels sprouts are a very popular member of this family. On account of their size many people who do not like to serve poor, common old cabbage will serve these. Brussels sprouts are interesting in their growth. The plant stalk runs skyward. At the top, umbrella like, is a close head of leaves, but this is not what we eat. Shaded by the umbrella and packed all along the stalk are delicious little cabbages or sprouts. Like the rest of the family a rich soil is needed and plenty of water during the growing period. The seed should be planted in May, and the little plants transplanted into rich soil in late July. The rows should be eighteen inches apart, and the plants one foot apart in the rows.

"Kohlrabi is a go-between in the families of cabbage and turnip. It is sometimes called the turnip-root cabbage. Just above the ground the stem of this plant swells into a turnip-like vegetable. In the true turnip the swelling is underground, but like the cabbage, kohlrabi forms its edible part above ground. It is easy to grow. Only it should develop rapidly, otherwise the swelling gets woody, and so loses its good quality. Sow out as early as possible; or sow inside in March and transplant to the open. Plant in drills about two feet apart. Set the plants about one foot apart, or thin out to this distance. To plant one hundred feet of drill buy half an ounce of seed. Seed goes a long way, you see. Kohlrabi is served and prepared like turnip. It is a very satisfactory early crop.

"Before leaving the cabbage family I should like to say that the cabbage called Savoy is an excellent variety to try. It should always have an early planting under cover, say in February, and then be transplanted into open beds in March or April. If the land is poor where you are to grow cabbage, then by all means choose Savoy.

"Carrots are of two general kinds: those with long roots, and those with short roots. If long-rooted varieties are chosen, then the soil must be worked down to a depth of eighteen inches, surely. The shorter ones will do well in eight inches of well-worked sandy soil. Do not put carrot seed into freshly manured land. Another point in carrot culture is one concerning the thinning process. As the little seedlings come up you will doubtless find that they are much, much too close together. Wait a bit, thin a little at a time, so that young, tiny carrots may be used on the home table. These are the points to jot down about the culture of carrots. I am saying very little about depths and distances because these were all worked out by the boys last winter and tables may be had for the asking.

"Peter covered the ground of celery raising. One or two points only I will speak of. A very rich, workable soil is a celery soil. In the process of getting such a bed ready it ought to have a thorough wetting down a day before planting. Celery seed is small, and, as is the case with other small seed, it is quite likely to be planted in clumps or bunches. To avoid this the seed should be mixed with sand and then sprinkled in drills. These drills should be six inches apart, and very shallow. Sow the seed, cover, and water. This bed should be screened from direct sunlight. A careful, gentle, daily watering is necessary. Thin the little plants to four inches apart. Peter can tell the rest of the story. The only other thing I need add is this, that in transplanting the work should be done quickly. One should not take up celery plants and perhaps leave them a long time before placing in new quarters. Plant immediately. It takes little to upset a celery plant and check its growth. So never take up a whole lot of celery plants at one time. Take up a few, keep them moist, plant them, and then start again.

Elizabeth Sowing Small Seed From the Package. Photograph by Helen W. Cook.

"I am not going to say a word about corn. You all remember the boys' work of last year. Your teacher has planned next Friday afternoon for the boys to discuss this at school. So we shall not have our meeting, but I shall come to the school to hear their report. I hope our girls will take notes on this. For I know that Eloise has decided to raise corn next year; Helena is going to; and Leston will not be out of the corn contest this season.

"The cucumber is the next vegetable in the line. This is a plant from foreign lands. Some think that the cucumber is really a native of India. It is believed that it was brought here by the Negro, and that a species of cucumber from Africa became finally at home in this country. A light, sandy and rich soil is needed—I mean rich in the sense of richness in organic matter. When cucumbers are grown outdoors, as we are likely to grow them, they are planted in hills. Nowadays, they are grown in hothouses; they hang from the roof, and are a wonderful sight. In the greenhouse a hive of bees is kept so that cross-fertilization may go on.

"But if you intend to raise cucumbers follow these directions: Sow the seed inside, cover with one inch of rich soil. In a little space of six inches diameter, plant six seeds. Place like a bean seed with the germinating end in the soil. When all danger of frost is over, each set of six little plants, soil and all, should be planted in the open. Later, when danger of insect pests is over, thin out to three plants in a hill. The hills should be about four feet apart on all sides.

"Egg plant is another vegetable we have not tried. It is another of those which has been improved by crossing, usually with peppers and tomatoes. But as we are not Burbanks yet, I shall not talk of that side of egg plant culture. Some varieties of egg plant grow to a large size but the smaller fruits, on the whole, have the better flavour. A good, well-worked, rich garden loam is the soil for this vegetable. The seed may be planted out in the open in little drills six inches apart. The seed should be scattered along as lettuce seed is. When the plants are about six inches high, transplant them to their permanent place. They should then stand about two feet apart on all sides. More often the seeds are started inside in March. When the little plants are about two inches high they should be transplanted into boxes or pots. Screen from too hot sunshine. About the time of corn planting the plants should go into the open. A rich soil is now quite necessary. Again I would suggest as a good method the placing of a little well-rotted manure under each place where an egg plant is to go. There is a rather interesting parasite which sometimes fastens itself upon the egg plant. A parasite is a form which clings to another and takes its nourishment from this latter or host. The parasite is a lazy shirk. So in this case the parasite grows on the egg plant and absorbs the food which the egg plant needs for itself. This is not an uncommon thing in nature. When such a thing happens first pull up and destroy the poor egg plant, for the parasite is clinging too closely and persistently to be removed.

"Nearly everyone in our country clings to lettuce as the only plant for a salad. Over across the sea in old England this is not so. Other plants are used in this way and called salad plants. Endive is one of these. Some of the endives are curly-leaved, and when blanched are attractive to look upon; and surely there is no reason why we should not consider the side of beauty in vegetables.

"Endive is a very hardy plant. One-half an ounce of the seed will sow one hundred feet of drill. Sow it as you would lettuce seed in soil which should be moist and rich. The plants finally should be about eight inches apart in the drill. The outer leaves should be tied over the top in order to blanch the inner ones for table use. In the fall the plants from summer sowings may be taken up with balls of earth on the roots and placed in the coldframes ready for use through the winter. The coldframe is a blessing. It is a place of storage all the fall and early winter. It is a place for early work in the late winter and first part of spring.

"Lettuce with children has always been a prime favourite for cultivation. Before the time of Christ, lettuce was grown and served. There is a wild lettuce from which the cultivated probably came. There are a number of cultivated vegetables which have wild ancestors, carrots, turnips and lettuce being the most common among them. Lettuce may be tucked into the garden almost anywhere. It is surely one of the most decorative of vegetables. The compact head, the green of the leaves, the beauty of symmetry—all these are charming characteristics of lettuces.

"Not all form heads. There is a mistaken idea abroad among children that by transplanting, any lettuce can be made to head. Only such varieties as are called heading lettuces will head. And these must be transplanted in order to have really good heads. There are two general types of lettuce—the Cos and the cabbage. The cabbage grows more like a cabbage with great tendency toward heading. The Cos grows longer, narrower, and has spoon-shaped leaves, which have a big, coarse midrib. The inner leaves cling more closely together after a heading fashion; the outer leaves spread apart. We grow in our American gardens more of the cabbage type than of the Cos. Should we go to see our French cousins next summer, the Cos lettuce would be served to us with plenty of oil as a dressing.

"As the summer advances and as the early sowings of lettuce get old they tend to go to seed. Don't let them. Pull them up. None of us are likely to go into the seed-producing side of lettuce. What we are interested in is the raising of tender lettuce all the season. To have such lettuce in mid and late summer is possible only by frequent plantings of seed. If seed is planted every ten days or two weeks all summer, you can have tender lettuce all the season. When lettuce gets old it becomes bitter and tough.

"Melons are most interesting to experiment with. We suppose that melons originally came from Asia, and parts of Africa. Watermelons grow wild in Africa. The Negroes and wild animals feed upon them. Perhaps that is the reason why the coloured people so love them. Anyway, melons belong to these countries. Melons are a summer fruit. Over in England we find the muskmelons often grown under glass in hothouses. The vines are trained upward rather than allowed to lie prone. As the melons grow large in the hot, dry atmosphere, just the sort which is right for their growth, they become too heavy for the vine to hold up. So they are held by little bags of netting, just like a tennis net in size of mesh. The bags are supported on nails or pegs. It is a very pretty sight I can assure you. Over here usually we raise our melons outdoors. They are planted in hills. Eight seeds are placed two inches apart and an inch deep. The hills should have a four foot sweep on all sides; the watermelon hills ought to have an allowance of eight to ten feet. Make the soil for these hills very rich. As the little plants get sizeable—say about four inches in height—reduce the number of plants to two in a hill. Always in such work choose the very sturdiest plants to keep. Cut the others down close to or a little below the surface of the ground. Pulling up plants is a shocking way to get rid of them. I say shocking because the pull is likely to disturb the roots of the two remaining plants. When the melon plant has reached a length of a foot, pinch off the end of it. This pinch means this to the plant: just stop growing long, take time now to grow branches. Sand or lime sprinkled about the hills tends to keep bugs away.

"Onions are about as popular a vegetable as we have. Some people are quite scornful of onions because of their truly disagreeable odour. But I do not know what we should do without the onion for flavourings. Peter is to plant onions where he last had celery. That is very wise, because onions do especially well coming after a crop for which the land was heavily fertilized. Onions like moisture of soil, too. If the soil is not rich enough, nitrate of soda may be added. The most discouraging thing about chemical fertilizers is the fact that advertisements say to have a certain quantity for an acre of land. Few boys and girls are planting entire acres, to just one thing. Now, suppose you write down this: Add ¼ pound of nitrate of soda to 100 square feet of land then use the proper fractional amount. To buy 1/40 of a pound for example sounds absurd. Buy your quarter pound and put the approximate amount on. Sprinkle chemical fertilizer over the surface of the soil and rake it in just under the surface.

"There are two methods for the planting of onions. One way is to use seed; the other, sets. Sets mean little onion bulbs. These are placed in drills about six inches apart and so that the little bulb may be just beneath the surface of the soil. Do not set too low. These bulbs are ready before seed onions. Seeds are rather slow in development. If you make sowings pretty thick the tender tops may be used, and so the thinning process is done to advantage.

"I believe that all your gardens should have some parsley in them. It can be planted as a border, since it grows low and has a fringy, decorative effect. If you were Italian girls and boys you would have parsley if you had nothing else; for the Italians always use it in their soups. There is a European variety, not as pretty as the common variety, which grows taller and ranker. It has a stronger smell. An Italian boy who was living in America had a garden in which he grew both kinds of parsley. He was asked which he liked the better. Straightening up, he pointed to the European saying, 'Smells stronger.'

"I believe none of us have raised parsnips. It takes ¼ of an ounce of seed to plant 100 feet of drill. Any deep, rich, moist garden soil will do. Just as early as the ground is workable, the seed should be sown. Sow in drills of ½ inch depth and 6 inches apart. The plants should finally stand about eight inches apart. They may be dug before or after frost. Some people think that the early frosts improve the flavour of parsnips.

"I suppose there is no vegetable so well liked as peas. Who would wish a Fourth of July dinner without peas? The early varieties of peas go into the ground just as early as possible. I like best to dig trenches six inches deep and about eighteen inches apart. As the peas are dropped into the trench, cover over with about two inches of soil. As the plants grow, fill soil into the trench. Of course, peas have to be brushed. So as soon as the little runners form put brush behind the plant and start the twiners about the brush stalks. A variety of pea called Gradus is very excellent in flavour.

"As for potatoes—well, now, just see Peter grin! He has covered that subject. Of course, I can add nothing to an expert's advice.

"Peppers are worth trying. If you do not care for them in your home gardens, add them to the school garden. They work in with the courses in cooking. Just as egg plants are started inside, so ought peppers to be. Whenever the soil is warm and the weather settled, the pepper plants may go out. The best soil for them is a rich, sandy one. The little plants should have about one and a half foot of space on all sides. At first they look pretty lonesome so far apart but soon they will grow to large, bushy plants. A little hen manure mixed with soil and put on top of the ground about the little peppers gives them a good start in their new quarters. There are many interesting kinds of peppers to grow. If a pepper with a little sting is wished try such varieties as Bird's Eye, Red Cluster, and Tobasco. Suppose the peppers are to be used for stuffing. Then large, rather more mild-flavoured kinds are needed. Ruby King pepper is a bouncing beauty. The Red Etna, Improved Bull Nose and Golden King are other good ones.

"The word pumpkin stands for good, old-fashioned pies, for Thanksgiving, for grandmother's house. It really brings more to mind than the word squash. I suppose the squash is a bit more useful, when we think of the fine Hubbard, and the nice little crooked-necked summer squashes; but after all, I like to have more pumpkins. And as for Jack-o'-lanterns—why they positively demand pumpkins. In planting these, the same general directions hold good which were given for melons. And use these same for squash-planting, too. But do not plant the two cousins together, for they have a tendency to run together. Plant the pumpkins in between the hills of corn and let the squashes go in some other part of the garden.

"We have very nearly worn out the subject of radish. About the only cultural point I would add is this: Make radish develop quickly. If growth is slow, the radish is likely to be poor. Sometimes all the growth goes to top. Fine, green leaves result, but no good radishes. Then doctor the soil in order that fruit development may be quickened. Radishes are the quickest in maturing of all vegetables. That is why your teacher had radish beds for the lower grades in the school. The children got a result and got it quickly. Josephine might have raised radishes as well as parsley in the box garden she had. People in cities could raise these two vegetables on their roofs just as well as not. They are worth the effort it takes in trying.

"There is one point I wish to speak of in tomato culture. The great trouble in cultivating comes in the overgrowth of vines. Each plant becomes a large, overgrown, unwieldy sort of affair unless looked out for. Use a stake for each vine. Tie the main stalk of the plant to this. Let the development of fruit come from the top of the plant. So pinch back the lateral branches and remove these. In this way the tomato garden is a neat and pretty one. This treatment is similar to that given grapes.

"There is a tomato called the dwarf champion. This is a dwarf variety and so gives less trouble than the other kinds. It does not get troublesome and often does not need staking. If you were little boys and girls, I should say plant this kind of tomato every time.

"I have little more to say about turnips. They are an old vegetable, for over two thousand years ago, the Greeks dined on turnips. I usually plant mine right out in the open. The soil may be a sandy one.

X

FLOWER CULTURE

In planting the flower garden there are a few things always to be considered. These are matters of colour, of time and of persistency. "The subject of colour is not so trying, after all, as many gardeners seem to think. If you people wish to plant a few of a good many gay-blooming plants, then I guess colour is a problem. The chief thought in a flower garden should be, how I can make a beautiful picture of this garden of mine. You see right off how tiring and dazzling the garden of too many little dots of colour could be. Look about in nature—see the beautiful range of the butterfly weed, the pinky purple of Joe Pye, the scarlet of cardinal flowers, the blue of certain asters, the pink of bouncing Bet, the yellow of tansy and goldenrod. Nature is constantly presenting perfect splashes of brilliant colour here and there. And yet it is not inharmonious. Why? One reason is that much of the colour is in great masses, and what is not has been softened by stretches of soft green.

"Let us take a hint from this for our small gardens. Plant colours in masses, and have breaks of green in between. Not a bad idea! I seem to hear you say.

"Then a garden should have a strong time element about it. By this I mean that one should plan a garden for a round of bloom. Why have all the blooms in August? If you look at this bulb time-table I shall have ready for you, you will find that it is possible to start with bulbs. Snowdrops and crocuses will gladly usher in a continuous round of bloom for you. I do not mean that these two will bloom continuously. Not at all! But I mean they are the starters. There are conditions, where spring bloom and fall bloom alone is desired. This is the case with all school gardens where summer care is impossible. Another table called a garden of continuous bloom will help you plan this.

"Another point to think of is persistency. Why not plant some seed which will produce plants that come up year after year? Why not have some hardy perennials and some self-sowing annuals? Poppy and cornflower sow themselves. These are annuals. Think of the perennials, which come year after year to welcome us. I think you should have hardy matter in your gardens. Peonies come up year after year, iris takes care of itself, helianthus or perennial sunflower bobs up each year.

"George asked me one day, what I meant when I spoke of herbaceous plants. A herbaceous plant is one with a non-woody stem, as geraniums. Mock orange is not herbaceous, because it has a woody stem. When I speak of hardy plants, I mean those which can stand living outdoors through their existence, from start to finish. A half-hardy plant is one that requires fostering before being planted outdoors. We consider asters half-hardy, because they need the extra heat for start which the nasturtium does not need. You would not think of starting nasturtiums indoors. But asters and stock really need this sort of a start in life.

"The tall flowers must go toward the back of the garden, for if they were placed in the foreground they would screen the others. The plants of medium height make up the main part of the garden; while the low plants are in the foreground as borders.

"Perhaps it would be wiser to put some tall perennials or self-sowing annuals in the background, and among the shrubbery. Then save bed spaces for the annuals. This will cause less disturbance in the garden than the sowing of annuals in with the perennials.

"I cannot take up all the garden flowers with you, because it is an impossibility. But a certain number of the more common ones I will talk of.

"There are certain plants rather easier to grow than others, and very satisfactory in results. One of these all-around plants is the pansy. It likes best of all cool, moist places but it will do well, under rather reverse conditions. Pansies are the easiest of plants to grow from seed, and they offer a ready response to experiments with cross-fertilization. The very best time to sow plants is after midsummer. Anyway, the work must be done before October the first. Let us claim then that the middle of August is a good time. Make little drills a quarter of an inch deep for the seed; or better, sprinkle it on a fine seed bed. Over the seed sift a little soil. Pansy seed is fine and small, so great care must be taken in the early waterings; better far to cover the bed with old sacking, and water the sacking. In this way, the seed is not washed away. The little sturdy plants should be covered over with leaves or straw for the wintertime. When early spring comes, you will be delighted with plants which are well along.

"Pansy flowers should always be kept well picked. Do not let the seed pods form if you desire continuous bloom. It is well to hold this in mind—that if plants are hurried along too fast, the flowers suffer in size. Small, inferior flowers result from such treatment. Pansies have a habit of running out—that is, the flowers grow smaller each year. It is merely a warning to keep making new sowings in order that one may always have large, vigorous blooms.

"Choice seed of this flower is very expensive. It is a plant that some florists have put all their time upon. It has seemed to certain men that one of the greatest things in the world, is to find out ways of improving the plants of the earth. So certain fruits are crossed to make new and better ones; and certain flowers are being constantly worked over to get superior strains. Sweet peas, pansies, stock, and dahlias are plants which have been much improved by man's skill.

"Larkspur is one of those plants which children so rarely try. I have wondered often why. It is not hard to raise, and so I am hoping that some of you will try it another season. The larkspur is a hardy plant, and there are both annuals and perennials in this family. Some varieties are dwarfed, and grow only a foot and a half high, while others grow five feet in height. This latter growth is very charming in the background of the garden. The flower spikes are showy and the foliage pretty. The larkspur likes a pretty rich sort of soil. The seed is very slow in germinating, and that is reason enough for fall planting. The stay over winter gives these fussy seeds time to make up their minds to germinate. This sowing should be done after the middle of October.

"Really charming blue flowers are a bit difficult to find because we have fewer blue flowers than those of the reds and yellows. Do not get the impression that larkspurs are only blue in colour. There are yellow, pink, red, and white varieties. But the blue is very fine. So when you are thinking of high flowers for backgrounds, keep the larkspur in mind.

"Hollyhock is another good background plant, because of its height and sentinel-like effect. It sows itself, so will take care of itself. Perennial phlox is well to put into the garden. Helianthus, I have mentioned, as suitable for backgrounds. It has a rather bad habit of too free spreading.

"Peonies are very satisfactory. I am sure you will all want some of them. They look their best planted in clumps. A certain pink and white peony is called the rose peony. It is sweet scented, and when in blossom it scents the portion of the yard where it is placed. These look well planted in wide borders. The roots, or bulbs, should go about three inches below ground in nice, rich, garden soil. Do not plant where they get the full blaze of early sun.

"I'd put some iris in the border, too. It requires no care. You need not bed it over, even, in the fall. It likes a certain amount of moisture, but grows readily under almost all conditions. The German iris is an easy grower; the French fleur-de-lis is lovely with its more delicate blossom. Certain irises, to be sure, are particular about their quarters, but the two kinds mentioned are not. They like a certain amount of open space. Do not hide them in the shrubbery, although they may be planted near it.

"You might put in some hardy chrysanthemums. These need good rich garden soil. They should also be placed near the back of the garden for good effect. You may choose almost any colour in these. Some of the little button chrysanthemums are good for backgrounds. The yellow ones make good splashes of colour, while the dull reds are most beautiful. These bloom after frost. When the frost has made havoc with the foliage, cut the plant down to about one inch of the ground. It is well to cut the flowers before frost.

"You have now a few good background plants which are hardy.

"The biennials can be so planted as to behave like perennials. These plants, you remember, are doing their best blossoming work the second summer. So by yearly sowings you may always have good effects. I have mentioned some already for your garden:— Canterbury bell, cornflower and foxglove are biennials. Cornflower tends to self sow, but needs help in this work from you. Sweet William is an old favourite. Of course, it is pretty gaudy. But I like old sweet William in spite of his gay tendency. They are rather stiff, but so easy to raise, being not very particular about anything.

"When it comes to annuals there is a multitude of these to plant each season. There are candytuft and alyssum for borders. Then mignonette is absolutely necessary to keep the garden sweet. Coreopsis is easy to raise, and so is godetia. If a great big bold mass of colour is desired, put in Shirley poppies. These grow well even on sandy soil. It is well to remember, that these do not lend themselves kindly to transplanting.

"Suppose there is a bit of sandy ground which needs a low-growing plant. Put in this spot portulaca. The bright little blossoms, constantly blooming, add a bit of cheer to that old sandy place.

"There are the old stand-bys which are good bloomers— nasturtiums, zinnias, marigolds and petunias. In the case of zinnia, it is better to buy these seeds by the ounce. Children's penny packages and the regular five-cent packages are filled usually with seeds which produce variously coloured blossoms. One can plan for no good effects in this way. If you get a seed catalogue, and look through the zinnia list, you can choose just what you like.

"Certain plants are spoken of as plants for bedding. These plants are placed in a formal bed after the spring flowers have finished their blooming. You sometimes see in the park fine beds of tulips and hyacinths early in the season. After these have finished their blooming, plants which are all started are put in their bed. If seeds were planted they would take so long to develop that the bed would look bad for a long time. So bedding plants are put in. Geraniums are the most popular of all. Begonias, fuchsias, heliotrope and coleus are often used. Geraniums will stand almost any kind of soil, and therefore have great advantages over most plants. Begonias will flourish in the shade; while the strong point about coleus is that of beauty of foliage.

"To those of you who have started outdoor bulb beds, the bedding list will be of some service. Marguerites look well in such a bed. Often one sees a border of ageratum about such a one. There is always a sort of stiff effect about such borders, however. A canna bed is after the same order, yet is effective. Salvia, or scarlet sage, looks well in wide borders, or near the underpinning of the house. Both these may act as bedding plants.

"There are three other kinds of gardens I should like to bring to your minds—the rock garden, the herb garden and the wild-flower garden. This last we shall have to leave for another time, however.

"Whenever a rockery is mentioned to some people they shrug their shoulders, and murmur something about a mere heap of rocks. Now, a rock garden may be very pretty, or very ugly. Such a garden should never be stuck out in the front yard to hit one in the face. But if you have a place in your yard, which is near the woods or in the vicinity of trees, or by a rocky ledge—in short, if you have any place with a bit of wildness surrounding it, use this for a rockery. If your yard is just a plain, tame, civilized yard, you'd better leave the rock garden out. I know of a lady living in a city, whose backyard is a rocky ledge. That ledge itself told her what her garden ought to be. It just cried out to her, 'Build a rock garden on me.' And she did it. Any other kind of a garden would have been out of place and taste there. Wherever a rocky ledge is found, there is a possibility for a rock garden.

"To have a good garden of this sort, one must have earth as well as rocks. Earth must be put into all the crevices of rock, so that there is some depth to it, and at such an angle that it won't be washed out by hard rains. A rock garden should have an earth foundation. I mean that there must be much of earth about it. I saw a charming one, which had only climbing nasturtiums planted over it. It was a great rock jutting out, and extending back into the yard—a big, flat, irregular affair—and all over it were these running vines. It was very simple and very effective. Go to the woods and seek out ferns which are growing in rocky places. Take what little earth they have about them, and try to give them a similar position in your own rockery. Bring back some leaf mould from the woods, and mix the garden soil for the rockery. Candytuft, dwarf phlox, stonecrop, morning glory, saxifrage, bleeding heart, rock cress, myrtle, thrift, columbine, bell flower, and moss pink. Get some moss, too, for chinks between rocks.

"If we could go back to old colonial days, and visit a dame's garden, I am sure we should find a little herb garden there. Our mothers might call these herbs pot herbs. Here all the flavourings for the soups were raised. Here sweet lavender might be found, its flowers used to make fragrant the bed linen. Horehound, anise and others were used in medicines; while little caraway seeds made delicious the cakes and cookies. I can see bunches of dried sage hung in the attic.

"Even with us there might be good use made of this garden both at home and at school. We do, of course, grow parsley, which is an herb, but the others seem to have dropped out of our gardens. We might at least grow next summer the sage and savoury for the turkey stuffing.

"Herbs need a sandy, well-worked soil. Seed should be sown in drills about twelve inches apart. The seed should be sown in early spring, as soon as the ground is warm. Sprinkle the seed just below the surface, and cover lightly with soil.

"A list of common herbs includes the following: Anise, balm, basil, borage, caraway, catnip, coriander, dill, fennel, horehound, hop, hyssop, lavender, pot marigold, sweet and pot marjoram, parsley, pennyroyal, rosemary, rue, sage, savoury, tansy, sorrel, thyme, and wormwood. It would be of little use to plant all of these, even to see what the plants were like. I would suggest your trying lavender, sage, savoury, and dill.

"Lavender seed is very slow to germinate, so sow the seed plentifully in early spring. The soil should have a dusting of lime over it as lavender plants enjoy lime. The flower is the part you wish. Pick these flower stalks before the flowers get old. Dry, and then sprinkle the dried flowers in the linen chest. Lavender is very sweet, and is often spoken of as sweet lavender. To this day one will hear women singing in the streets of London, 'Sweet lavender, buy my sweet lavender.'

"Sage likes a good, well-drained soil. It, too, likes lime. The little seedlings should be thinned out to stand about ten inches apart. When you see flowers forming cut the sage plant and quickly dry. It makes a pretty border plant in the garden. Savoury is also a border plant. But this is a hardy annual, while lavender and sage are perennials. It likes a light but rich soil. Both the leaves and flowers are used in soup flavouring. "Dill is also sown in early spring. It is the seed of the dill plant, and not the leaves and flowers, which is the useful part. The seeds are used in the making of pickles.

"I shall hope to see something in the herb line, in your gardens next year—a hardy garden started, and a good bit of taste displayed by all of you. You girls might raise mint to put in lemonade.

"Next time we shall have our talk on wild flowers. Some of you know and love many of these wild flowers."

ANNUALS FOR SANDY SOILS

NAME	COLOUR	TIME	HEIGHT	SPECIAL POINTS
Clarkia	Rose White Purple	June Oct.	1½ ft.	Good to use for a border plant. Use single varieties.
Poppy	White Red	July Aug.	1½ ft.	Do not transplant. Fine for mass effects.
Nasturtium	Yellow to Maroon	June Oct.	1-5 ft.	Most satisfactory, especially for cut flowers. Blooms freely.

Portulaca	White Red Yellow	June	½ ft.	Blooms freely — grows close to the ground.
Zinnia	Red Magenta	June Frost	1½ ft.	Grows without great care. Blooms freely. Looks best when massed.

ANNUALS FOR HEAVY SOILS

NAME	COLOUR	TIME	HEIGHT	SPECIAL POINTS
Godetia	Red White	July Oct.	1-2 ft.	Easy to grow. An English favourite: blooms freely.
Sweet Pea	Variety	July Oct.	4-6 ft.	Plant early in the spring in a sunny spot. Keep flowers well clipped for constant bloom.
Petunia	Magenta	May	1-2 ft.	Mass or use in borders. Give plenty of sun. Good for the outdoor window box.
Sweet Alyssum	White	July Sept.	½ ft.	Sow thickly in the borders. Blooms freely. Grows in every soil.

Pot Marigold	Orange Yellow	June Frost	1½ ft.	Blooms freely. Looks well in masses. Stiff effect as a cut flower.

SEVEN FAVOURITE ANNUALS FOR CUT FLOWERS

NAME	COLOUR	TIME	HEIGHT	SPECIAL POINTS
Sweet Alyssum	White Red	July Sept.	½ ft.	Use in low bowls for table decorations.
China Aster	White Blue Purple	July to Oct.	1 ft.	Lasts long after cutting.
Baby's Breath	White Rose	June Oct.	2-3 ft.	Use in bouquets with other flowers.
Coreopsis	Yellow Brown	June Nov.	1-2 ft.	Place by themselves in tall vases.
Nasturtium	Scarlet Yellow	June Oct.	1-5 ft.	Cut freely for constant bloom. Use in great masses in low vases.
Pansy	Purple Blue Yellow White	May June	½-1 ft.	Cut closely. Place in low dishes.
Pink	White Maroon Rose	May Aug.	1 ft.	Lasts well after cutting.

ANNUALS THAT BLOOM AFTER FROST

NAME	COLOUR	TIME	HEIGHT	SPECIAL POINTS
Sweet Alyssum	White	July Sept.	½ ft.	Withstands early frosts. Good border plant.
Candytuft	Red White	June to Sept.	½ ft.	Sow at intervals through the summer. Good cut flower.
Cornflower	Rose Blue White	June	2-3 ft.	Good for cut flowers. Blooms freely.
Marigold	Yellow Browns	June Aug.	½-2 ft.	Better for garden effects than for cut flowers.
Annual Phlox	Yellow Reds White	May Aug.	½-1 ft.	Self sows. Good all round plant.
Ten-weeks stock	White Purple Pink	June July	1½ ft.	A second sowing made in May flowers the same season.

FRAGRANT ANNUALS

NAME	COLOUR	TIME	HEIGHT	SPECIAL POINTS
Mignonette	Green	May Oct.	1 ft.	Beautifully fragrant. Do not transplant. Good to use as a break for

				clashing colours.
Sweet Peas	White Scarlet Blue Yellow	July Oct.	4-6 ft.	Free bloomer. Try variety *Lathyrus odorata.*
Ten-weeks stock	Pink Purple White	June July	1 ft.	Good for cut flowers. Fragrant at night.
Sweet Sultan	White Yellow Purple	June Aug.	2 ft.	*Centaurea moschata* good for cut flowers.
Sweet Alyssum	White	July Sept.	½ ft.	*Alyssum maritimun.* Low growing, border plant.

ANNUALS THAT RE-SOW THEMSELVES

NAME	COLOUR	TIME	HEIGHT	SPECIAL POINTS
Morning glory	Reds Blues Whites	July Oct.	15-30 ft.	Grows rapidly. Makes a good screen.
Poppy	Pink Scarlet White	June Sept.	½-2 ft.	*Papaver Rhocas* and *P. somniferum.*
Phlox	Yellow Reds White	May Aug.	½-1 ft.	*Phlox Drummondii.* Need much water.
Pot Marigold	Orange Yellow	June Oct.	1-2 ft.	*Calendula officinalis* Likes a warm soil.

ANNUALS THAT CLIMB

NAME	COLOUR	TIME	HEIGHT	SPECIAL POINTS
Balloon Vine	White	Aug.	10 ft.	Grows rapidly. Good screen.
Japanese Hop	Inconspicuous	July Oct.	8-20 ft.	Rapid grower. Looks well growing along old fences.
Moon-flower	White	July Sept.	15-30 ft.	Night bloomer. Grows rapidly.
Morning Glory	Purple White	June Aug.	20 ft.	Rapid grower. Good screen.

ANNUALS FOR SHADY PLACES

NAME	COLOUR	TIME	HEIGHT	SPECIAL POINTS
Godetia	White Red	July to Oct.	1-2 ft.	The flowers are showy.
Musk	Yellow	June to Aug	½-1 ft.	Need moisture and coolness.
Nemophila	Blue White	June to Oct.	1-3 ft.	Moisture, partial shade and coolness.
Pansy	White Yellow Purple	May to Oct.	½-1 ft.	Sow under same conditions as musk in early spring. Blooms freely in the fall.

ANNUALS FOR SUNNY PLACES

NAME	COLOUR	TIME	HEIGHT	SPECIAL POINTS
Balsam	Red Yellow White	June to Oct.	1-2 ft.	Plant in rich sandy loam under direct rays of sun.
Cornflower	Blue	May to Sept.	2 ft.	Full flower. Resows itself
Cosmos	White Pink Crimson	July to Sept.	4 ft.	Overrich soil retards bloom.
Gaillardia	Red Yellow	July to Oct.	3 ft.	Good cut flo-wer. Blossoms freely.
Marigold, Afri-can	Yellow	July to Frost	3 ft.	Blooms profu-sely; stiff flo-wer head.
Nasturtium	Yellow Scarlet	June to Oct.	1-5 ft.	Both dwarf and tall varie-ties are rapid growersand free bloomers.
Rose moss	White Magenta	June Aug.	½-¾ ft.	*Portulaca grandiflora.* Plant in posi-tion direct sunshine.
Verbena	Various	July to Sept.	2 ft.	Start inside for early bloom.
Wallflower	Orange	June	1½ ft.	Sown in Sept.

		July		blooms in May.

ANNUALS FOR ROCKY PLACES

NAME	COLOUR	TIME	HEIGHT	SPECIAL POINTS
Phlox	Red White	June to Oct.	1-1½ ft.	Variety *Phlox Drumondii* Duration of bloom depends on richness of soil.
Candytuft	Red White	May Frost	½-1½ ft.	Plant in the fall and cover for early spring bloom.
Clarkia	Purple Red White	June Oct.	1½ ft.	Use *Clarkia elegans.* Thrives in both sun and partial shade.
Nasturtium (Dwarf)	Reds Yellows	June Oct.	1 ft.	*Tropoeolum minor;* blooms very early.

FRAGRANT PERENNIALS

NAME	COLOUR	TIME	HEIGHT	SPECIAL POINTS
Winter Heliotrope	Lilac	Dec. Feb.	1 ft.	Blooms outdoors in winter. Flowers small.
Russian Violets	Violet	March	½ ft.	Double flowers. Hardy.
Lily-of-the-	White	May	¾ ft.	Plant by mi-

Valley				ddle of March for that season's bloom. Needs part shade. Spreads.
Valerian	Pinkish	June	3 ft.	Finely cut foliage. Easy to grow.
Lemon Lily	Yellow	June	2 ft.	Flowers 4 in. long. Tubers which multiply rapidly.
Fringed Pink	Lilac	July	1 ft.	Blooms until autumn if prevented from seeding.
Bee Balm	Scarlet	July Aug.	2 ft.	Odour of mint. Good for mass effects.
White Day Lily	White	Aug. Sept.	2 ft.	Lilies 4-6 in. long. Fine, broad-leaved foliage.

PERENNIALS FOR CUT FLOWERS

NAME	COLOUR	TIME	HEIGHT	SPECIAL POINTS
Christmas Rose	White	Nov. Feb.	½ ft.	Blooms outdoors in the snow. No fragrance. Charming in masses.
California Vio-	Violet	March	½ ft.	Large flowers

lets				but single. Fragrant.
Foxglove	Purple	June	3-4 ft.	Large flowers, long stems.
Oriental Lark-spur	Blue White	June	2-3 ft.	Beautiful colour, long stems. Let no flowers go to seed.
Japan Iris	Variety of	July	3-4 ft.	Short lived when cut. Fine blooms.
Japan Anemone	Pink White	Sept.	3 ft.	Finest September flower. Plant in spring. Plant for afternoon sun.

TALL PERENNIALS

NAME	COLOUR	TIME	HEIGHT	SPECIAL POINTS
Hollyhock	Variety of	July	6 ft.	Single varieties are the hardier; double varieties last longer.
Plume Poppy	Pinkish	July	6 ft.	Spreads rapidly. Fine for massing and screening. Do not plant on the west as it shuts off sun.
Golden Glow	Yellow	Aug.	6 ft.	Multiplies

				rapidly. Fine bloomer. Liable to pest lice. Spray with soap solution.
Double Sunflower	Yellow	Aug.	5 ft.	Largest double flower of any perennial. Likely to run out unless divided
Late Sunflower	Yellow	Sept.	10-12 ft.	Tallest of perennials; blooms till October.

LOW GROWING PERENNIALS

NAME	COLOUR	TIME	HEIGHT	SPECIAL POINTS
Crested Dwarf Iris	Blue	April	9 in.	Earliest of iris. Good for edgings. Plant the middle of March.
Dwarf Flag	Purple	April	9 in.	Increases rapidly. Large flowers. Good colouring
English Daisy	Pink White	April	6 in.	Good for spring budding purposes.
Tufted Pansies or Violets	Blue Yellow White	June	6 in.	More but smaller flowers than pansies. After

				July cut back, manure and they will bloom again in September.
Carpathian Harebell	Blue	July	8 in.	Bloom for 6 weeks. Easy to grow.
Coral Bells	Red	Aug.	12 in.	Grow in sandy, well-drained soil. Coral red flowers.
Popp-mallow	Crimson	Sept.	9 in.	Blooms nearly 12 weeks. Colour does not harmonize with others.

PERENNIALS OF MEDIUM HEIGHT

NAME	COLOUR	TIME	HEIGHT	SPECIAL POINTS
Bleeding Heart	Pink	May	1½ ft.	Long lived and long of bloom. Graceful.
European Peony	Crimson White	May	3 ft.	Earliest of peonies. Poor appearance in the fall.
Sweet William	Red Pink White	June	12 in.	Self sows. Flowers at their best the second year.
Chinese Peony	Crimson	June	2½ ft.	Long-lived.

	Pink White			Very satisfactory. Plant in September.
Foxglove	Purple White	June	3-3¼ ft.	Spire-like cluster of flowers.
Oriental Larkspur	Blue	June	2-3 ft.	Best blue perennial. Cut flower spikes as soon as they fade.
Oriental Poppy	Red	June	3 ft.	Self sows. Flowers 6 in. across.
Gaillardia	Red Yellow	June Nov.	1 ft.	Flowers more freely than any other perennial. Cover plants after ground freezes.
Late Phlox	All best Blue and Yellow	Aug. Sept.	1½ ft.	Fragrant in the evening. Many colours of bloom.
Hardy Chrysanthemum	Blue Scarlet	Sept. Nov.	3ft.	Long season of bloom. Deep rich soil and sunny exposure for best results.

XI

THE WILD-FLOWER GARDEN

"A wild-flower garden has a most attractive sound. One thinks of long tramps in the woods, collecting material, and then of the fun in fixing up a real for sure wild garden.

"If the wild garden is to be a school affair, then I certainly should plant the different kinds of flowers together. The north corner near the building is a suitable place. But if the garden is to be at home—your own private little garden—I am inclined to think it would be better to plant the wild flowers here and there among the cultivated ones.

"A wild-flower garden is a joy each year, because up it comes without constant replanting of seed. It is a hardy garden. As Nature often covers her wood-flowers over with leaves preparatory to winter, so you might copy her and do the same.

"Many people say they have no luck at all with such a garden. It is not a question of luck, but a question of understanding, for wild flowers are like people and each has its personality. What a plant has been accustomed to in Nature it desires always. In fact, when removed from its own sort of living conditions, it sickens and dies. That is enough to tell us that we should copy Nature herself. Suppose you are hunting wild flowers. As you choose certain flowers from the woods, notice the soil they are in, the place, conditions, the surroundings, and the neighbours.

"Suppose you find dog-tooth violets and wind-flowers growing near together. Then place them so in your own new garden. Suppose you find a certain violet enjoying an open situation; then it should always have the same. You see the point, do you not? If you wish wild flowers to grow in a tame garden make them feel at home. Cheat them into almost believing that they are still in their native haunts.

"Wild flowers ought to be transplanted after blossoming time is over. Take a trowel and a basket into the woods with you. As you take up a few, a columbine, or a hepatica, be sure to take with the roots some of the plant's own soil, which must be packed about it when replanted.

"The bed into which these plants are to go should be prepared carefully before this trip of yours. Surely you do not wish to bring those plants back to wait over a day or night before planting. They should go into new quarters at once. The bed needs soil from the woods, deep and rich and full of leaf mold. The under drainage system should be excellent. Then plants are not to go into water-logged ground. Some people think that all wood plants should have a soil saturated with water. But the woods themselves are not water-logged. It may be that you will need to dig your garden up very deeply and put some stone in the bottom. Over this the top soil should go. And on top, where the top soil once was, put a new layer of the rich soil you brought from the woods.

"Before planting water the soil well. Then as you make places for the plants put into each hole some of the soil which belongs to the plant which is to be put there.

"I think it would be a rather nice plan to have a wild-flower garden giving a succession of bloom from early spring to late fall; so let us start off with March, the hepatica, spring beauty and saxifrage. Then comes April bearing in its arms the beautiful columbine, the tiny bluets and wild geranium. For May there are the dog-tooth violet and the wood anemone, false Solomon's seal, Jack-in-the-pulpit, wake robin, bloodroot and violets. June will give the bell-flower, mullein, bee balm and foxglove. I would choose the gay butterfly weed for July. Let turtle head, aster, Joe Pye weed, and Queen Anne's lace make the rest of the season brilliant until frost.

"Let us have a bit about the likes and dislikes of these plants. After you are once started you'll keep on adding to this wild-flower list.

"There is no one who doesn't love the hepatica. Before the spring has really decided to come, this little flower pokes its head up and puts all else to shame. Tucked under a covering of dry leaves the blossoms wait for a ray of warm sunshine to bring them out. The last year's leaves stay on through the winter brooding over the little fresh sprouts. These embryo flowers are further protected by a fuzzy covering. This reminds one of a similar protective covering which new fern leaves have. In the spring a hepatica plant wastes no time on getting a new suit of leaves. It makes its old ones do until the blossom has had its day. Then the new leaves, started to be sure before this, have a chance. These delayed, are ready to help out next season. You will find hepaticas growing in clusters, sort of family groups. They are likely to be found in rather open places in the woods. The soil is found to be rich and loose. So these should go only in partly shaded places and under good soil conditions. If planted with other woods specimens give them the benefit of a rather exposed position, that they may catch the early spring sunshine. I should cover hepaticas over with a light litter of leaves in the fall. During the last days of February, unless the weather is extreme take this leaf covering away. You'll find the hepatica blossoms all ready to poke up their heads.

"The spring beauty hardly allows the hepatica to get ahead of her. With a white flower which has dainty tracings of pink, a thin, wiry stem, and narrow, grass-like leaves, this spring flower cannot be mistaken. You will find spring beauties growing in great patches in rather open places. Plant a number of the roots and allow the sun good opportunity to get at them. For this plant loves the sun.

"The other March flower mentioned is the saxifrage. This belongs in quite a different sort of environment. It is a plant which grows in dry and rocky places. Often one will find it in chinks of rock. There is an old tale to the effect that the saxifrage roots twine about rocks and work their way into them so that the rock itself splits. Anyway, it is a rock garden plant. I have found it in dry, sandy places right on the borders of a big rock. It has white flower clusters borne on hairy stems.

"The columbine is another plant that is quite likely to be found in rocky places. Standing below a ledge and looking up, one sees nestled here and there in rocky crevices one plant or more of columbine. The nodding red heads bob on wiry, slender stems. The roots do not strike deeply into the soil; in fact, often the soil hardly covers them. Now, just because the columbine has little soil, it does not signify that it is indifferent to the soil conditions. For it always has lived, and always should live, under good drainage conditions. I wonder if it has struck you, how really hygienic plants are? Plenty of fresh air, proper drainage, and good food are fundamentals with plants.

"It is evident from study of these plants how easy it is to find out what plants like. After studying their feelings, then do not make the mistake of huddling them all together under poor drainage conditions.

"I always have a feeling of personal affection for the bluets. When they come I always feel that now things are beginning to settle down outdoors. They start with rich, lovely, little delicate blue blossoms. As June gets hotter and hotter their colour fades a bit, until at times they look quite worn and white. Some people call them Quaker ladies, others innocence. Under any name they are charming. They grow in colonies, sometimes in sunny fields, sometimes by the road-side. From this we learn that they are more particular about the open sunlight than about the soil.

"If you desire a flower to pick and use for bouquets, then the wild geranium is not your flower. It droops very quickly after picking and almost immediately drops its petals. But the purplish flowers are showy, and the leaves, while rather coarse, are deeply cut. This latter effect gives a certain boldness to the plant that is rather attractive. The plant is found in rather moist, partly shaded portions of the woods. I like this plant in the garden. It adds good colour and permanent colour as long as blooming time lasts, since there is no object in picking it.

"I suppose little children would not have a perfect spring without the dog's tooth violet. The leaves are attractive and almost make the beauty of a bouquet. It is sometimes called trout lily. The mottled effect of the leaves accounts for the trout part of the name, and as for lily, it *is* a lily, and never belonged to the violet family at all. Dig the plant up, and the bulbous root tells the story. It really does belong to the lily family. The nodding yellow flower is pretty, too. These, when picked, last a long time in water. They like to grow in the neighbourhood of the brook. A moist, half-shaded half-open piece of land is their delight, and therefore in many gardens the trout lily might have to be left out.

"There is a sweet little flower called the wood anemone, or wind-flower. It is another modest little flower, white in colour. The constant nodding of the petals stirred by even a breath of wind gives it the name of wind-flower. These also grow in colonies. Have you noticed how social, but clannish, our wild flowers are? Especially is this true of the real woods flowers, rather than of the wayside flowers. The anemone grows in open places by the woods or the hillside. They are a sort of border plant evidently trying to leave the woods, but still bound to it.

"If in your yard there happens to be a big old fatherly tree or a decaying stump, plant wind-flowers all about it. You may make the flowers feel that they are on the edge of the woods.

"While I have numbered bloodroot among May flowers, it often does appear in April, and before the wood anemone. The silvery, white blossom pushes its head above the leaves in a fine fashion. They are sensitive flowers, closing partly in cloudy weather, and actually dropping to pieces in a rainstorm or under severe winds. The leaves are large, rather coarse, but pretty with their light under surfaces. The stems have tinges of red on them, a dark red sap in the roots. These roots bleed when disturbed. The Indians used to stain their faces with this orange sap-blood. You will find bloodroot growing in rich soil either in open woods or on rocky slopes.

"In a nice, rich, moist place put a few Jack-in-the pulpits. This flower is much like a child's jack-in-the-box. It is so different from most of our plants that it has the effect of the joker in a pack of cards. Push back the flap over Jack's face and you will see a club like a policeman's billy. Along this club the inconspicuous flowers are borne. Later, in the fall, the fruit forms, and inside, instead of rather uninteresting flowers, are bright red berries. So Jack jokes again.

"There is always a great feeling of joy when the first trilliums, or wake robins, appear. Walking in the deep, moist woods suddenly one sees a mass of big leaves and white flowers. The same irresistably lovely trilliums have come again. Three big leaves, then a flower stalk shooting up from the centre of this whorl of leaves, and on top the crowning glory—the three-petaled trillium flower. A fragrant white or pink form is called the nodding wake robin. These in a glance tell their wishes. The plant sometimes is nearly two feet high. So a clump of these could easily go toward the back of the wild-flower garden in shade and moist soil.

"Another wild flower of striking beauty is the May apple or mandrake. It comes very early in May, often in April. This plant grows to about the same height as the trillium. Only the big spreading leaves of the mandrake are visible at first sight. Beneath these, and daintily hung in the junction of the leaf stalks, is the lovely, waxy, white blossom. Late after the fading of the blossom the fruit appears. So its name of May apple comes from this fruit, which has a sickly sweet taste. The leaf and stalk part of the May apple are of a poisonous nature. This flower, too, likes rather low, moist, shaded places.

"The false Solomon's seal is found in woods where moisture is. During June and July this plant is in blossom. After the white flowers the fruit, or berry, appears. The berry changes from green, to white, to red. There is a two-leaved Solomon's seal called the false lily-of-the-valley which is found at this same time. It has usually two little lily-like leaves and a blossom stalk running up from these. Tiny fragrant flowers are borne on this stalk. These plants grow in moist woods, also. One might plant these two near together in the garden, for the soil conditions are the same for both.

"Who would wish a wild-flower garden without violets? The little sweet wood ones, the big horse shoes, the rare white, and more rare yellow—any and all are worth our while! Violets, at least the most of them, prefer not to be huddled away. I wonder why, when people think of transplanting violets, a dull, dark, moist spot immediately comes to mind? Violets like the sun, like good soil, and plenty of air. Some violets are found in the swamps, but did you happen to notice what long stems they have? Why? The reason is to raise the lovely flowers into the light. Nothing could be sweeter or more satisfactory than a violet bed. I rather like violets bedded by themselves. They fill in corners beautifully. They grow gladly about trees. They adorn borders. You may cover them, in the fall or not as you like. They are not fussy. Take a north corner at school, a corner not wholly shaded by any means—fill that in solid with violet plants in the fall. That corner always will be a thing of real beauty.

"The bellflower coming in May blooms on until September. The flower is blue, purple or violet. It is a flower found in dry places, on grassy slopes, along hillsides, and is common to most localities.

"I have a sneaking fondness for mullein. One or two stalks of it give a charming effect in the garden. Its yellow flowers, its tall flower stalk, the thick, hairy leaves—all these are its charms. It is said that these same hairy leaves were used as wicks by the ancients. Anyway, the flowers themselves on the tall stalks that often reach to seven feet, look like gleaming lights on a torch. The mullein has a simple dignity. It grows in the dry fields and along roadsides. So you see it is by no means particular about its habitat, its place of abode.

"Another tall plant is the foxglove. The flowers are gathered together in a sort of spike at the end of the stalk, are large and yellow and really lovely. The plant grows to about four feet in height. It has a bad habit, this downy false foxglove, of absorbing some of its nourishment from the roots of plants near which it stands. This plant, too, is fond of dry places.

"A very gay flower, intensely red, is the bee balm. It is an herb, and a perennial. It is often called Oswego tea, because the Indians are supposed to have used it for tea. Then, again, you will hear it called Indian's plume. This name seems most suitable. I can just imagine a chief strutting around with this gay plume on his head. It likes a somewhat secluded, moist, shady, cool place. I think it would be possible for some of you to make it grow at home. For colour it would be invaluable. The cardinal flower is the only flower more gaudy in red than this bee balm.

"When one comes to orange colour the butterfly weed takes the prize. This flower has a variety of names: it is called pleurisy root, and wind root, and orange root. Would you think that this gay little beggar was a member of the milkweed family? It is. When seed time comes it produces a seed pod like unto the milkweed pod only more slender than this. All summer long the insects hover about it. It is just like a signal to them. "Come over here to me!" it calls to them all. It is found in dry places, in the fields and pastures, along the dusty road sides, and by the sooty railroad track it flashes its signal.

You can make this plant feel at home surely. And think of the butterflies that will visit your garden all summer long.

"Then later comes old Joe Pye weed. Joe Pye was an Indian doctor but that doesn't seem to have anything to do with his weed. Yes, it has its connection. For when old Joe Pye went out on a case of typhoid fever he carried this plant along; hence, its name. The plant sometimes grows to ten feet in height. Really the swamp is its home. So if you are to use it at all remember that it must have this condition of great moisture, even to swampiness. The flower clusters are of a charming colour, a beautiful dull pink.

"Another inhabitant of wet places is the turtle head. The flower resembles in shape a turtle's or a snake's head, and so receives both names.

"When it comes to Queen Anne's lace, you say that is a troublesome weed. Yes, it is. But it is truly beautiful with its lacy flower head. A great bouquet of these on the porch, the dining table, or the school piano is a real picture. A clump of these in the garden, if held in check, is simply stunning. How can they be held down? The only way is to let no flower heads go to seed. The little, clinging, persistent, numerous seeds are seeds of trouble. This lovely bother grows in any sort of soil.

"There are numbers and numbers of wild flowers I might have suggested. These I have mentioned were not given for the purpose of a flower guide, but with just one end in view—your understanding of how to study soil conditions for the work of starting a wild-flower garden.

"If you fear results, take but one or two flowers and study just what you select. Having mastered, or better, become acquainted with a few, add more another year to your garden. I think you will love your wild garden best of all before you are through with it. It is a real study, you see."

XII

LANDSCAPE GARDENING

The subject to-night is a very pretentious one, for no one would expect boys and girls to be landscape gardeners. But many boys and girls have excellent taste and taste is the foundation stone of landscape gardening. This work has often been likened to the painting of a picture. Your art-work teacher has doubtless told you that a good picture should have a point of chief interest, and the rest of the points simply go to make more beautiful the central idea, or to form a fine setting for it. Look at that picture over Miriam's head. See that lone pine, the beautiful curve of the hillside, the scrub undergrowth about the tree, the bit of sky beyond! As soon as one looks at that picture one's eye rests on the pine, and the other features seem to appear afterward.

"So in landscape gardening there must be in the gardener's mind a picture of what he desires the whole to be when he completes his work. Take, for example, your school grounds. You did a bit of landscape work there, although we never called it that before. The little schoolhouse itself was our centre of interest. How could we fix up the grounds so that the little building should have a really attractive setting? That, I believe, was the thought in each of your heads, although no one of you ever put this into words.

"Notice now with me the good points about that work, and from this study we shall be able to work out a little theory of landscape gardening.

"First there is a good extent of lawn about the building, the path to the door is slightly curved and pleasingly so, a fine little maple stands out rather interestingly on the side lawn, the flower garden has a good mass effect, the screen of poplar trees at the back acts as a stately rear guard, and the vines over the outbuilding hide what was once a blemish.

"Let us go back to the lawn. A good extent of open lawn space is always beautiful. It is restful. It adds a feeling of space to even small grounds. So we might generalize and say that it is well to keep open lawn spaces. If one covers his lawn space with many trees, with little flower beds here and there, the general effect is choppy and fussy. It is a bit like an over-dressed person. One's grounds lose all individuality thus treated. A single tree or a small group is not a bad arrangement on the lawn. Do not centre the tree or trees. Let them drop a bit into the background. Make a pleasing side feature of them. In choosing trees one must keep in mind a number of things. You should not choose an overpowering tree; the tree should be one of good shape, with something interesting about its bark, leaves, flowers or fruit. While the poplar is a rapid grower, it sheds its leaves early and so is left standing, bare and ugly, before the fall is old. Mind you, there are places where a row or double row of Lombardy poplars is very effective. But I think you'll agree with me that one lone poplar is not. The catalpa is quite lovely by itself. Its leaves are broad, its flowers attractive, the seed pods which cling to the tree until away into the winter, add a bit of picturesqueness. The bright berries of the ash, the brilliant foliage of the sugar maple, the blossoms of the tulip tree, the bark of the white birch, and the leaves of the copper beech—all these are beauty points to consider.

"Place makes a difference in the selection of a tree. Suppose the lower portion of the grounds is a bit low and moist, then the spot is ideal for a willow. Don't group trees together which look awkward. I never should have Peter and Myron march together in school. Why? Because they look wretchedly together. Myron makes Peter look short and Peter causes Myron to look overgrown. So it is with trees. A long-looking poplar does not go with a nice rather rounded little tulip tree. A juniper, so neat and prim, would look silly beside a spreading chestnut. One must keep proportion and suitability in mind.

"I'd never advise the planting of a group of evergreens close to a house, and in the front yard. The effect is very gloomy indeed. Houses thus surrounded are overcapped by such trees and are not only gloomy to live in, but truly unhealthful. The chief requisite inside a house is sunlight and plenty of it.

"There are no shrubs on the school grounds. You had spoken of doing that but bulbs took up the attention of the girls this fall. And as for you boys—you were attending to your own crops. Shrubbery is very pleasing if properly placed. It is just the thing to fill in corners near buildings, to help define the turns in walks, and to use as hedges. Usually one shrub standing by itself is not nearly so pleasing as one tree by itself. It has a squatty and isolated appearance. There is a corner close by the school building where shrubs should go. Why? Because the place looks bare and staring, and the building is very ugly at that point; the shrubs would fill in the space, and make the building look much better.

"As trees are chosen because of certain good points, so shrubs should be. In a clump I should wish some which bloomed early, some which bloomed late, some for the beauty of their fall foliage, some for the colour of their bark and others for the fruit. Some spireas and the forsythia bloom early. The red bark of the dogwood makes a bit of colour all winter, and the red berries of the barberry cling to the shrub well into the winter. This list of shrubs which Philip has made out will be a help to you in this work.

PHILIP'S SHRUB TABLE

COMMON NAME	BOTANICAL NAME	HEIGHT	COLOUR	SPECIAL POINTS
		March		
Spice Bush	*Benzoin odoriferum*	6-15 ft.	Yellow	Flowers appear before leaves. Crimson fruit in fall. Aromatic odour.
Daphne	*Daphne Mezereum*	4ft.	Purple	The only hardy deciduous daphne. Plant in light soil and in shade.
		April		
Barberry	*Berberis-Thunbergii*	2-4 ft.	Yellow	Prefers dry soil. Berries all

COMMON NAME	BOTANICAL NAME	HEIGHT	COLOUR	SPECIAL POINTS
				winter.
Golden Bell	*Forsythia suspensa*	5-8 ft.	Yellow	Flowers appear before leaves. Hardy; free from insects.
		May		
Red-osier Dogwood	*Cornus stolonifera*	4-8 ft.	White	Red branched. Plant in moist soil.
Japanese Snow Flower	*Deutzia gracilis*	1-3 ft.	White	Very beautiful when flowering. Needs well drained soil.
Japanese Snowball	*Viburnum plicatum*	8 ft.	White	Not as likely to have lice as common snowball. Larger balls.
Lilac	*Syringa vulgaris*	15 ft.	Purple	Very fragrant. Will grow anywhere even in some shade.
		June		
Deutzia	*Deutzia Lemoinei*	1-3 ft.	White	Hardy; flowers showy.
Weigela	*Diervilla Florida*	6 ft.	Pink White Red	May have white or red flowers. Flo-

COMMON NAME	BOTANICAL NAME	HEIGHT	COLOUR	SPECIAL POINTS
				wers under trees. Lives where other shrubs die.
Spirea	*Spiræa Van Houttei*	4 ft.	White	Most showy of spireas. Grows anywhere.
Mock Orange	*Philadelphus Coronarius*	10 ft.		Varieties of different colours. Fragrant; makes good screen.
Smoke Bush	*Rhus cotinus*	4-10 ft.	Purplish	Hardy. Beautiful all summer. Purple colour changes to smoke colour.
		July		
Spirea	*Spirea Bumalda*, var. *Anthony Waterer*	3 ft.	White	Flowers run from white to deep pink. Late flowering. Hardy.
Sweet Pepper Bush	*Clethra alnifolia*	3-10 ft.	White	Moist soil or sandy. Late blooming; fragrant flowers.
		August		

COMMON NAME	BOTANICAL NAME	HEIGHT	COLOUR	SPECIAL POINTS
Althea, Rose of Sharon	*Hibiscus Syriacus*	12 ft.	White to purple	Very hardy. Plant in any good garden soil.
		September		
Hardy Hydrangea	*Hydrangea paniculata*	8 ft.	White to pink	A showy shrub. Flowers remain on all winter.
		October		
Witch Hazel	*Hamamelis Virginiana*	6-20 ft.	Yellow	Grows anywhere. Likes moisture. Fruit "explodes."

"Certain shrubs are good to use for hedge purposes. A hedge is rather prettier usually than a fence. The Californian privet is excellent for this purpose. Osage orange, Japan barberry, buckthorn, Japan quince, and Van Houtte's spirea are other shrubs which make good hedges.

"You have to remember that not only should grounds look well to the passerby but they should look equally well from the inside of the building. As your mother is working in the kitchen during the hot summer or sewing during a long dull winter afternoon, would it not be a joy to her to look out at a syringa sweet with blossom or a barberry with nodding red berries? Landscape gardening is not only for the purpose of adding beauty to the earth's surface, but also for the putting joy into the heart of a person as well.

"I forgot to say that in tree and shrub selection it is usually better to choose those of the locality one lives in. Unusual and foreign plants do less well, and often harmonize but poorly with their new setting.

"I spoke of the path to the schoolhouse with its slight curve. Landscape gardening may follow along very formal lines or along informal lines. The first would have straight paths, straight rows in stiff beds, everything, as the name tells, perfectly formal. The other method is, of course, the exact opposite. There are danger points in each.

"The formal arrangement is likely to look too stiff; the informal, too fussy, too wiggly. As far as paths go, keep this in mind, that a path should always lead somewhere. That is its business—to direct one to a definite place. Now, straight, even paths are not unpleasing if the effect is to be that of a formal garden. The danger in the curved path is an abrupt curve, a whirligig effect. It is far better for you to stick to straight paths unless you can make a really beautiful curve. No one can tell you how to do this.

"Garden paths may be of gravel, of dirt, or of grass. One sees grass paths in some very lovely gardens. I doubt, however, if they would serve as well in your small gardens. Your garden areas are so limited that they should be re-spaded each season, and the grass paths are a great bother in this work. Of course, a gravel path makes a fine appearance, but again you may not have gravel at your command. It is possible for any of you to dig out the path for two feet. Then put in six inches of stone or clinker. Over this, pack in the dirt, rounding it slightly toward the centre of the path. There should never be depressions through the central part of paths, since these form convenient places for water to stand. The under layer of stone makes a natural drainage system.

"A building often needs the help of vines or flowers or both to tie it to the grounds in such a way as to form a harmonious whole. Vines lend themselves well to this work. It is better to plant a perennial vine, and so let it form a permanent part of your landscape scheme. The Virginia creeper, wistaria, honeysuckle, a climbing rose, the clematis and trumpet vine are all most satisfactory.

"Just close your eyes and picture a house of natural colour, that mellow gray of the weathered shingles. Now add to this old house a purple wistaria. Can you see the beauty of it? I shall not forget soon a rather ugly corner of my childhood home, where the dining room and kitchen met. Just there climbing over, and falling over a trellis was a trumpet vine. It made beautiful an awkward angle, an ugly bit of carpenter work.

"Of course, the morning-glory is an annual vine, as is the moon-vine and wild cucumber. Now, these have their special function. For often, especially in school work, it is necessary to cover an ugly thing for just a time, until the better things and better times come. The annual is 'the chap' for this work.

"Along an old fence a hop vine is a thing of beauty. One might try to rival the woods' landscape work. For often one sees festooned from one rotted tree to another the ampelopsis vine.

"Flowers may well go along the side of the building, or bordering a walk. In general, though, keep the front lawn space open and unbroken by beds. What lovelier in early spring than a bed of daffodils close to the house? Hyacinths and tulips, too, form a blaze of glory. These are little or no bother, and start the spring aright. One may make of some bulbs an exception to the rule of unbroken front lawn. Snowdrops and crocuses planted through the lawn are beautiful. They do not disturb the general effect, but just blend with the whole. One expert bulb gardener says to take a basketful of bulbs in the fall, walk about your grounds, and just drop bulbs out here and there. Wherever the bulbs drop, plant them. Such small bulbs as those we plant in lawns should be in groups of four to six. Daffodils may be thus planted, too. You all remember the grape hyacinths that grow all through Katharine's side yard.

"The place for a flower garden is generally at the side or rear of the house. The backyard garden is a lovely idea, is it not? Who wishes to leave a beautiful looking front yard, turn the corner of a house, and find a dump heap? Not I. The flower garden may be laid out formally in neat little beds, or it may be more of a careless, hit-or-miss sort. Both have their good points. Great masses of bloom are attractive.

"You should have in mind some notion of the blending of colour. Nature appears not to consider this at all, and still gets wondrous effects. This is because of the tremendous amount of her perfect background of green, and the limitlessness of her space, while we are confined at the best to relatively small areas. So we should endeavour not to blind people's eyes with clashes of colours which do not at close range blend well. In order to break up extremes of colours you can always use masses of white flowers, or something like mignonette, which is in effect green.

"The old-fashioned flowers are lovely—sweet William, phlox, old-fashioned pinks, petunia, verbena, zinnia, marigold, mignonette, and poppy are always dear and sweet. Hollyhocks are charming. They represent a kind of guard for the garden. Stand this hollyhock phalanx up against a wall like naughty boys, close to the house, or by an old fence. They are so tall that they must be in the background. They grace it. Otherwise they would overtop and shadow the other garden plants. If there is an old ash pile, an old dump or anything else unsightly, plant something tall before it. Hollyhocks would not do for this, since their foliage is too scanty. Castor beans are just the thing, however; and sunflowers, the old giant ones, are good, too. A screen is for screening, so that the foliage is of first consideration.

"A wild-flower garden is a good scheme, too. What is lovelier? Bank in a north corner full of these. Hepatica, columbines, anenome, bellflower, butterfly weed, turtle head and aster represent wild flowers which bloom from March through October. I can see that north corner now. Miriam has planned to have one, and has really done the work this fall.

"The water garden is another good thing to try with just the right setting. A place at the end of a slope of land, near some drooping trees, a bit shaded would be right. The garden Philip made is a pattern for you all to follow.

"Finally, let us sum up our landscape lesson. The grounds are a setting for the house or buildings. Open, free lawn spaces, a tree or a proper group well placed, flowers which do not clutter up the front yard, groups of shrubbery—these are points to be remembered. The paths should lead somewhere, and be either straight or well curved. If one starts with a formal garden, one should not mix the informal with it before the work is done.

"At one time we said a little about poor taste in garden furnishings. Painted kettles, old drain pipes, whitewashed bricks, and edgings of shells seemed to us then a bit fussy and crude. So, too, is a summer house stuck out on the front lawn, a rustic seat all by itself in an open spot, an archway which forms an arch over nothing. The summer house should be placed in the side yard, or in the rear in a spot where trees lend it a background. If its use is that of a resting spot for your mother, she certainly would not wish it right out on the front lawn. If the house is for children to play in, then again it is not for the front of the house. An appropriate place is near the garden where it makes a cool place to rest after labour, a spot from which to view the beauties of the garden, and a charming place to serve afternoon tea.

"A good general plan to follow in this landscape work is to see what natural charms your place has, and then try to increase and help these. 'Help Nature' is a good watchword. Even though the garden plan is to be a formal one, the natural resources and setting of your place should be kept in mind. The little we did last year on the school grounds was a bit of landscape garden work. I did not call it that to you then, for if I had you would have been scared off. Philip's work in his backyard was of the same nature. The girls' flower garden was a bit of formal work. I guess, too, the outdoor bulb planting which Albert scorned might come under the same head. So you see you have been landscape gardeners without knowing it. To continue to be, all we have to do is to go on somewhat along the general lines I have spoken of to-night. Different committees have prepared a number of tables which should help you much in matters of selection."

GARDEN OF CONSTANT BLOOM BY MONTHS

NAME	COLOUR	HEIGHT	SPECIAL POINTS
		March	
Columbine	Red	1 ft.	Grows on rocky places. Graceful flower.
Hepatica	White Blue Pink	6 in.	Early spring flower. Ready to blossom under the snow. Last year's leaves shelter

			flower.
Saxifrage	White	8 in.	Grows in rocky, sandy places.
	April		
Bluebell	Blue	16 in.	Likes rock soil and sun.
Dwarf Iris	Blue	1 ft.	A good border plant. Does not require any special soil. Spreads.
Moss Pink	Pink	6 in.	Likes full sun. Spreads rapidly.
Violet	Blue	6 in.	Good soil. Plant in either sun or shade.
	May		
Lily-of-the-Valley	White	9 in.	Grows under trees, spreads rapidly. Flowers fragrant. Cut flower effect.
German Iris	Different	2 ft.	The best of flags for general planting purposes.
Forget-me-not	Blue	low	Thrives on moist soil. Planted with tulips follows them in bloom.
Chinese Peony	Different	4 ft.	The earliest of peonies. Good in borders.
Myrtle	Blue	low	Grows even in shade and poorly drained soils. Spreads rapidly.
	June		
Bleeding Heart	Pink	2 ft.	A hardy plant. Needs moist, good soil. Good border plant.
Foxglove	Purple	4 ft.	Perennial which self sows. Effective in backgrounds.

			Likes shade.
Garden Peony	Crimson	3 ft.	The real old-fashioned peony. Good border plant. Large blossoms.
Larkspur	Blue	4 ft.	Good for borders and backgrounds. The finest of blue flowers.
Sweet William	Different	2 ft.	A self sewing perennial. Bright colours. Good for massing. Sweet and constant bloomer.
		July	
Baby's Breath	White	3 ft.	Grows in rocky soil. Use for formal bouquets.
Butterfly Weed	Orange	2 ft.	Likes full sunlight and dry soil. Fine colour effect.
Perennial Phlox	Different	2-5 ft.	Good for borders and cut flowers.
Spiked Loosestrife	Pink	3 ft.	Belongs in wet swamp lands. Will grow in borders.
Hollyhock	Different	6 ft.	Use for backgrounds and borders.
		August	
Aster, New England	Blue	4 ft.	Grows in any soil. The best of tall asters.
Golden Glow	Yellow	6 ft.	Grows in any soil and spreads rapidly. Good background.
Japanese Clematis	White	Climber	Rapid flowering vine. Use on trellis. Sweet flowers.
Sunflower	Yellow	6 ft.	Fine for backgrounds and

			screens. Any dry soil.
Turtle Head	Rose Purple	2 ft.	Flowers on spikes. Any soil, but wet preferred.
	September		
Hardy Hydrangea	White	6 ft.	Blooms till frost. Blossom heads effective.
Japanese Anemone	Carmine	3 ft.	Good border plant. Blossoms last till frost.

WATER AND BOG PLANTS

COMMON NAME	SCIENTIFIC NAME	SPECIAL POINTS
Arrowhead	*Sagittaria latifolia*	One of the most popular water plants. Spreads badly.
Arum (water)	*Calla palustris*	Popular water plant. Grows less than 1 ft. high. Blooms in June.
Blue Flag	*Iris versicolor*	Grows from 2-3 ft. high. Grows in bogs and gardens.
Cat-tail	*Typha latifolia*	Grows to 8 ft. or more. Spreads rapidly.
Floating Heart	*Limnanthemum nymphoides*	Grows less than 1 ft. high. Good plant for a pond. Spreads readily.
Forget-me-not	*Myosotis Palustris*	Grows less than 1 ft. high. Prefers half-shady places.
Lotus, American	*Nelumbo lutea*	Good for a pond.
Marsh Marigold	*Caltha palustris*	Grows 1½ ft. high. Blooms in May.

Pickerel Weed	*Pontederia cordata*	Grows 3-4 ft. high. Blooms in July.
Pitcher Plant	*Darlingtonia Californica*	Grows less than 1 ft. Good for bog planting.
Sweet Flag	*Acorus Calamus*	Height is 2 ft. Flowers in early summer.
Water-lily	*Nymphaea odorata*	Sweet-scented, most popular water-lily.
Water Mint	*Mentha Aquatica*	One of the popular mint family. Low growing.

FOUR WATER-LILIES FOR BEGINNERS

NAME	COLOUR	SPECIAL POINTS
Nymphaea Gladstoniana	White	This is a hardy variety
Nymphaea Marliacea	Yellow	This is a hardy variety
Nymphaea dentata	White	Tender, night blooming plant
Nymphaea Zanzibariensis	Blue	Tender, day blooming plant.

WATER-LILIES FOR SMALL PONDS

NAME	COLOUR	REQUIRED DEPTH OF WATER
Nymphoea alba (hardy)	White	More than 2 ft.
Nymphoea tuberosa (hardy)	White	More than 2 ft.
Nymphoea Marliacea rosea (hardy)	Pink	More than 2 ft.
Nymphoea odorata, var. *minor* (hardy)	White	Less than 1 ft.
Nymphoea tetragona (hardy)	Yellow	Less than 1 ft.
Nymphoea Laydekeri, var. *rosea* (hardy)	Pink	Less than 1 ft.
Nymphoea Zanzibariensis (tender)	White	Less than 1 ft.

NOTE: – Any of these forms may be grown in from 1 to 2 ft. of water.

TREE TABLE

NAME	HEIGHT	SPECIAL POINTS
Carolina Poplar	100 ft.	Grows in a dry soil. Fastest growing street tree. Its dropping fruit is a nuisance. Sheds leaves early.
Catalpa	50ft.	Lovely white blossoms in June. Seed pods stay on into winter. Quick growing. Good lawn tree.
English Hawthorn	30ft.	Flowers in June. Red berries. Grows on dry soils. Slow grower. Sharp thorns.
Linden	90ft.	Easy to grow. Fragrant flowers. Rapid grower. European species smaller than American.

Live Oak	100 ft.	Not hardy in the North. Grows south of Virginia. Beautiful evergreen oak. Likes moist soil.
Locust	80ft.	Fragrant flowers in May and June. Rapid grower. Seeds in pods. Thorny bark.
Lombardy Poplar	90ft.	Quick grower. Stiff, straight and tall. Dignified but melancholy tree. Fine for pathway effect.
Norway Maple	100 ft.	Tall, well rounded tree. Yellow foliage in the fall.
Pin Oak	100 ft.	Fastest grower among oaks. You cannot grow plants under it.
Red Maple	100 ft.	Earliest flowering maple. Good for lowlands. Bright red foliage in the fall.
Sugar Maple	100 ft.	Moist soil. Bright foliage in the fall. Best street tree among maples.
Horse Chestnut	60 ft.	Fine white flowers in June. Attractive buds and leaves. Foliage grows very dense.

XIII

HOW BOYS AND GIRLS CAN MAKE MONEY FROM THEIR GARDENS

Naturally, we are all interested in ways and means of earning money. It is not a bad thing at all for a boy or girl to wish to turn work into cash. Not always is it possible for one to find a market next door for products. No, it is rarely as easy a matter as that. One has to really work a bit.

"Let me tell you one boy's story. This lad, let us call him Newton, had a nice vegetable and flower garden. He had worked so hard over it, it did seem to him as if he ought to be able to sell some of his produce. One day he loaded a little cart with vegetables and went down the street to a corner market. I imagine he went in a half-hearted sort of way. The market-man was busy and he spoke a bit roughly to the boy. But Newton went on to another store. He received the same sort of treatment there. This time he gave up discouraged and went home. His mother was not discouraged. She showed him how he should have made his vegetables, wagon and all, look more attractive.

"So Newton went to work again. He scrubbed his radishes and new carrots until they shone. He bunched them up into neat little bundles. Then the lettuce came in for its washing and cleaning. Thus he treated all the vegetables. Then he printed a sign 'Fresh Vegetables For Sale' and started off again. This time he went to the largest hotel in the little city in which he lived. There he was sent to the cook. This big, good-natured fellow said that he would look at his stuff. 'Looks good to me,' said the cook, 'it really looks like home-grown things,' Straightway he bought a good part of what Newton had and there and then made arrangements for daily deliveries of certain vegetables.

"The lesson from Newton's experience is this: in order to sell, you must put your wares in attractive shape. Who wishes to buy dirty radishes or droopy looking lettuce? No one is willing to pay decent prices. Putting materials in such condition that all the good points speak loudly at first, is one way to attract notice and sell later. If you find you can sell by shipping your goods the same points hold true.

"Another way to make money is to raise young plants for sale. Jack did this with his aster plants. Lots of people wish their garden plants partly started. They either do not have the interest, or else they have not the time for initial work. Asters, stock, tomatoes, lettuce, cabbage, pepper, celery—all of these may be started for market purposes.

"Suppose you have planted tomato seed. You are bound to have more young plants than you wish. Why not sell them? Suppose Mrs. Jones always buys hers. Then go to her and ask if she will not buy of you. She may not believe you can be a very good gardener, so she hesitates. Well, then just ask her if you cannot bring your little plants around for her to see when the time comes. Get to work in your best style. Transplant in little paper cups or strawberry baskets. Then the setting out of the plants will be very easy and quite a scientific performance. I think you will sell to Mrs. Jones all right.

"If you really intend to go into this early market side then you should certainly have a coldframe. You could not blame your mother if she refused to have the kitchen littered up with old tin cans and boxes all the spring. Do not be a nuisance at home just to make money.

Photograph by W.H. Jenkins

Myron Transplanting his Long-rooted Strawberry Plants.

Photograph by Helen W. Cooke

Katherine Transplanting Her Flowers by a Method of Lifting.

"I know a little girl who raised aster and stock plants, also young vegetable plants. She had a coldframe. In the spring, when people were starting their gardens and wondering where they should go for plants, she fixed up an attractive basket filled with her plants. She asked no exorbitant price, but a fair one for a little girl's good work. One year she bought herself a dressing table from her garden earnings. I think that well worth while. Don't you?

"Another way to make money from your garden is to sell your seed. I do not think any of you will be at all likely to try to rival the seed houses. But I am sure that you can supply certain seeds for your own fathers and mothers.

"Such seed as those of radish, lettuce and turnip you would not save. It is better to buy them. But surely you can make some pretty good selections for seed corn. I believe you can manage beans, peas, melons, pumpkins, potatoes and squash. Then we have, I believe, learned from the school flower garden how to select seed. Nasturtium seed may always be saved, dried and put into its own envelope. This will be found to be true, that seeds saved from our own flower garden often do not give satisfactory results as time goes on. The plants and flowers after a few seasons seem to spindle out. In the large seed gardens the varieties of flowers raised are either many or cross-pollination is carried on.

"In putting up your seeds in envelopes give a few cultural directions on them; that is, tell how to plant the special seed. Also, put on when to plant and the proper soil conditions under which planting should go on. If the seeds are those of flowers add information concerning height, colour of blossom, and time of blossoming. Someone might like to know also if the seed was that of an annual, biennial or perennial plant. Think out a neat, attractive way to fix these envelopes. If you do not wish to sell them, they will serve as nice Christmas gifts.

"Among the garden trappings which we made last winter are things you could easily sell. Such a plant stand as Jack made for his own room is certainly marketable. Make samples of your wares and then take orders for them. Again, these represent Christmas gifts, too.

"Rustic seats, a woven mat of corn husks to kneel on when weeding, a bit of nice trellis work, a little tool house are all possible pieces of work.

"I saw once what a boy called his handy boxes. These were wooden boxes, with hinged covers and handles, so that they might be carried about. One was for seeds. This box had partitions inside, and all the different envelopes of seeds were arranged in the different cubby holes. Another box had garden accessories. The word sounds interesting. It means all the little extras needed in the work. Labels, small stakes, a garden reel, measure, knife, cord, note book, pencil—all were in the box, all were things which the boy often used. You can make variations on these. But a box which may be carried about has advantages over one that is screwed up in the tool house.

"I believe the flower-gathering basket would sell well. It is not that it is a rather picturesque sort of Englishy custom to go out and pick flowers with a pretty basket tucked under one's arm, but it is very inconvenient, very hot work, and very mussy, to have to hold bunches of flowers in the hand as one gathers.

"In some places where there are summer colonies it is possible to sell bunches of flowers. I knew of a case where big bunches of sweet peas were brought to the hotel every morning. These sold for ten and fifteen cents the bunch and went like hot cakes.

"The girls may think of all sorts of wicker mats and trays that would make the garden tea more attractive. One ought to think of the aesthetic side.

"I have not mentioned working for others. Hire yourself out. Let it be known that you can and will weed, mow lawns, plant and transplant for so much per hour. Someone may be going off for a few weeks; see to it that you are the boy or girl to be employed. Prove yourself faithful.

"In the winter make garden utensils and also attend to the bulb end of it. At Christmas time you could do a big business.

"Someone might make and bottle kerosene emulsion. Paste on each bottle directions for using. Print very neatly, so it will look well.

"There are doubtless many other ways of making money. But, above all, do not neglect the other side; give away some things from your garden and some of your labour, too. If all you think of is the making of money the soul and heart of you all will get as small and shrivelled as a dry pea. Who wants to be stingy? Better never to make money than to grow like that. Don't let people pay you for everything you do. Do certain things for mother and father for nothing. The home garden is as much theirs as yours. Wouldn't it be ludicrous if your mother said, 'No, Katharine, you cannot have those flowers to carry to school unless you pay ten cents for them,' How cross you would be! Just as absurd, is it not, for you to suggest that you cannot work on that same garden unless you receive ten cents an hour? No, that is all wrong. And if any one of you feels that way do one of two things—either sit down and be ashamed for a good, long time and think of all the things done freely for you; or else go take all the money in your own little bank at home, buy something your mother wants, and give it, being glad, so glad you can get rid of what you have been so stingy about.

"Give flowers to the poor, the sick at home and the sick in hospitals, the church, the people you love, the people you think you don't love, and the people who seem lonely and forlorn.

"Once upon a time there was made a wondrous garden. It was called the earth. The flowers, the trees, the plants which afterwards became through man's skill our staple products—all these were free, absolutely free.

"If this is a true story, how can we be so small as always to make money from this garden? Let us pay our debt to it freely and gladly.

"This is our last talk. Some of you already have started your early vegetables and flowers. Instead of one coldframe we have four in our family and one belongs to a girl.

"It is going to be a better year of gardening than before. Leston is with us now. Another season there will be others. The school grounds look well, and if you have noticed the entire village looks a little better than ever before.

"We will shake hands all around. In a few weeks we shall have hands quite dirty with good old garden soil. You may take your stools and benches off with you, or leave them all here."

"We shall leave them," said Eloise; "for I am coming back often to sit on my little cricket right on your hearth."

"I am a little large for a cricket," went on Albert; "but I'd not quit this hearthstone, so my stool stays."

"And mine, too," each one added.

Off they trooped again, some down the country road, some up the road, others across the fields, and George, as usual, on his old horse. They shouted until out of sight.

"The best things in the world," the man murmured as he stepped out into the open and drew into his lungs deep breaths of the fresh spring air.